Renew by phone
0845
www.bristol.go
Bristol Libraries

D0635011

**PLEASE RETURN BOOK BY LAST DATE STAMPED**

# Praise for *Metamorphosis*

mportant and illuminating … Ryan writes with verve and
conviction, skilfully marshalling his facts and painting vivid pen-
portraits of his central characters – not least the engaging and
self-deprecating Don Williamson, 89, retired and relishing the
debate his theory has sparked.… A lively and thoroughly enter-
aining read; one of the best scientific books of recent years."

*New Internationalist*

"The ideas that Frank Ryan describes – fusion between quite
different creatures – at first seem ludicrous, but what great ideas
in science do not? … *Metamorphosis* has joined my shortlist of biological must-reads."

**Colin Tudge**, author of *The Secret Life of Trees*

"Ryan frames metamorphosis as one of the most mysterious of
all biological processes. He describes the role in elucidating it of a
colourful cast of characters including Jean-Henri Fabre, the 19th-
century French naturalist who studied silk moth pheromones,
and Vincent B. Wigglesworth, who pinned down the hormone
triggers of metamorphosis through decapitation experiments on
blood-sucking bugs …. A must for … the downright curious."

*New Scientist*

"Metamorphosis is the stuff of myth … Fascinating work."

*Times Higher Education*

"Has this not been *the* most exciting story in biology? This deeply
informed account delves into the fascinating history … It's a
saga, and Frank Ryan does it justice."

**Bernd Heinrich**, author of *The Nesting Season* and
*Mind of the Raven*

"Ryan brings an accessible passion to [his] subject comparable
to Carl Sagan's popularization of astronomy … *Metamorphosis*
ultimately leads the reader … to a call for greater understanding
and e

*rd Review*

Also by Frank Ryan

*Tuberculosis: The Greatest Story Never Told*

*Virus X*

*Darwin's Blind Spot*

*Virolution*

# Metamorphosis

## Unmasking the Mystery of How Life Transforms

Frank Ryan

ONEWORLD

A Oneworld Book

First published in trade paperback by
Oneworld Publications 2011
This paperback edition published 2012

ISBN 978–1–85168–913–2
ISBN 978–1–85168–862–3 (ebook)

Typeset by Jayvee, Trivandrum, India
Cover design by Richard Green
Printed and bound in Denmark by Nørhaven A/S

Oneworld Publications
185 Banbury Road, Oxford, OX2 7AR, England

*Metamorphosis: The action or process of changing in form or substance, especially by magic or witchcraft.*

The Shorter Oxford English Dictionary

*Metamorphosis: A profound change in form from [one] stage to the next in the life history of an organism.*

Webster's College Dictionary

# Contents

*Metamorphosis*

*Immature starfish (drawn by Haeckel)*

Prologue

# The Beautiful Mystery

IT IS A HOT DAY at the end of August and students at Amanda Elementary School in Manhattan, Kansas, stream out of classes to gaze heavenwards. After several days of cloud and rain, the weather has improved, and today the sun is beaming down out of a clear blue sky. Suddenly, fingers point to where splashes of golden-orange are soaring above the playground. Fluttering gaily, in a cloud drifting southwards, the great monarch butterfly migration has begun. Further south, in Emporia Village Elementary School, students are also staring up at the sky and counting. By the end of the day they will have already listed one hundred sightings. Later that afternoon, monarchs come pay a visit at Corinth Elementary School in Shawnee Mission. Hundreds float high in the air, then all of a sudden descend on the playground, accompanied by birds and dragon-flies that appear to be making the journey with them,

The monarch migration is an exciting annual attraction, when up to one hundred million butterflies migrate from the colder north to winter on the warm Californian coast or the Sierra Madre in Mexico. It is a journey as old as, and maybe a great deal older than, the great migrations of buffalo that have entered American legend. On the websites that link to the school biology lessons, teachers explain what is happening: the story of caterpillar and butterfly, two utterly different life forms following their different life cycles. The children are fascinated to learn of how the monarch 'changes its ecological niche entirely when it transforms from a caterpillar to a butterfly ... a miraculous biological process of transformation [involving] two ecologically different organisms, as distinct as a field mouse and a hummingbird'. It appears almost magical, a transformation that has captured the human imagination from classical times – and perhaps even longer.

We see a similar magic in the pictures that adorn rock shelters in Australia and South Africa, and deep within the gloom of French and Spanish caves, dating back to the Aurignacian period, roughly thirty thousand years ago. Painted by Stone-Age hunter-gatherers, they include rearing horses and charging rhinoceroses imbued with naturalistic grace. For scientists interested in human evolution, these earliest examples of figurative art are seen as a crucial threshold in the development of the creative imagination: they convey the quintessential qualities of what makes us human. Equally exquisite are sculptures dating from the same period, such as the figurines of a diving cormorant, or the putative head of a horse, carved out of mammoth ivory, which were discovered by the German palaeoanthropologist Nicholas Conard at the Hohle Cave in the Ache Valley.[1] One such ivory figurine, just a few centimetres long, depicts the hybrid features of a man and a lion. We can't be sure what the artist intended, but given that a second lion-man sculpture has been discovered in a nearby German cave, Conard suggests this might fit with a common shamanistic religious experience, noting 'the transformation between man and animal, and particularly between man and felines, was part of the Aurignacian system of beliefs'.[2] For Conard, this extension of creativity, from mere observation of nature, to a more ideational, perhaps spiritual inspiration, was the most exciting discovery of all. 'I've been digging for a long time and I'm usually very calm in my work – but this certainly got my heart pumping'. Anthropologist Christopher Chippendale, of the University of Cambridge Museum of Archaeology and Anthropology, has extensively documented that early cave art often features 'composite beings, from a world between humans and animals'. These, he says, '... are a common theme from the beginning of painting'.

There is a word for the dramatic transformation of one being into another: "metamorphosis", derived from the Greek *meta*, "with", as in to share in common, and *morph*, "form". This is the title Ovid chose for his classic collection of exotic and wonderful stories, so reminiscent of what was seen in the ancient sculptures and cave paintings, of how men and women lost their human forms and souls to be transformed, by divine miracle, into four-legged animals, birds, and flowering plants. It seems not unreasonable to assume such inspiration came from the everyday observation of the natural world, where the mystery of metamorphosis surrounds us. Indeed, the more one looks to nature, the more widespread is this phenomenon of bizarre, spectacular change. It extends far beyond the world of insects to include amphibians, such as the familiar frogs and toads, and in its most dazzling variety, to the strange and colourful marine invertebrate creatures that inhabit the hidden depths of the oceans, such as starfish, sea urchins, crabs, and sea squirts.

All such metamorphoses involve a series of transformations, so that the developing being exists in a variety of different stages, or forms, from the egg, through one or more "larval" stages, such as we see in the grubs or caterpillars of insects or the bewildering variety of marine invertebrate forms, to a transformation that results in the final, or "adult" form of the insect, or periwinkle, lobster, or salamander. An intriguing example is the startling metamorphosis of the starfish, *Luidia sarsi*, in which larva and adult co-exist simultaneously as independent life forms in the different ecologies of the surface waters and the ocean floor. If we did not know the larva and adult starfish were born from a single fertilised egg, we would view them as radically different animals belonging to completely different branches of the tree of life. In fact, this is no rare or fantastic freak: it is an altogether typical example drawn from the entrancing, some might say magical, process of transformation the starfish shares with the majority of the animal species that inhabit the oceans, lands, and air of our planet.

In nature, we are familiar with the metamorphoses of moths and butterflies, where the humble caterpillar stops its frantic feeding to enter a phase of quiescence, known as the pupa, or chrysalis. Walled off from the world, it appears to undergo a mysterious, and wholesale, transformation from which, as if by some quasi-miracle, the glory of the adult eventually emerges. For the Dutch anatomist Jan Swammerdam, judged by many to be the foremost naturalist of the seventeenth century, the metamorphosis of caterpillar to butterfly symbolised the journey from pedestrian life to death and resurrection in the afterlife, with the pupa representing the repose of the soul between death and the Day of Judgement. Such ecstatic vision is perhaps understandable when we see what emerges from this seemingly death-like cocoon, the newly emerging adult insect with its multi-faceted eyes, articulated legs, new-found sexual maturity, and extraordinary wings. That two such very different beings could derive from a single fertilised egg is at once shocking and thrilling. It is little wonder it has long intrigued the human imagination.

For a great thinker such as Aristotle, the butterfly's caterpillar stage was a continuation of its embryonic life before the formation of the perfect adult. 'The larva, while it is in growth', he argued, 'is nothing more than a soft egg'.[3] William Harvey, famous for his discovery of the circulation of the blood, saw some mystical influence at work in the transformation of the larva into an entirely new form. In the pages ahead we shall meet other naturalists who were equally enthralled, including Charles Darwin, the father of modern evolutionary theory, and the French naturalist and "poet" of nature, Jean-Henri Fabre. A modern example is Don Williamson, a marine biologist who has spent his working life at the former Port Erin Marine Laboratory in the beautiful setting of the Isle of Man. If Williamson is right, metamorphosis is even more intriguing than some of these past luminaries of science and philosophy dared to imagine.

Williamson will be seen as an iconoclast, a breaker of the conventional image of evolution. In one of his published papers, he explains the dilemma. 'The concept of "self" is quite sufficiently complicated in animals like ourselves, which do not metamorphose from one group to another during development', he writes, 'but it seems important ... to consider forms which, apparently, do undergo such changes and which add another dimension to the problem of identity'. In part, this book is his story, the encapsulation of a scientist inspired by a radical new theory, his struggle for credibility and the broader enlightenment his theory offers to science.

Williamson's ideas have provoked controversy in the fields of marine biology and entomology, so much so that some colleagues will likely object to my discussing his theory at all. By the conclusion of this book, I hope my readers will understand why I include him, in part because his story encapsulates the conflict with orthodoxy, the search for new areas of research as well as enlightenment. Metamorphosis is truly complex and profound, and thus no theory, whether orthodox or radical, should stand alone. So, in addition to important contributions from Darwin and Fabre, we shall focus on the revolutionary work of Sir Vincent Wigglesworth, widely regarded as the father of insect physiology; the American entomologist Carroll Williams; and the modern inheritors of both Wigglesworth and Williams's pioneering enlightenment, the husband-and-wife team of Lynn Riddiford and James Truman.

We cannot claim science has all of the answers – metamorphosis still jealously guards many of its secrets. But what we have captured to date is so precious and fascinating, it is no exaggeration to say that, even in its scientific exploration, metamorphosis remains both awesome and beautiful.

We shall examine these various scientific explorations of the mystery, which includes transformations in real life that appear every bit as exotic and strange as those we encounter in Ovid's fantasies. Metamorphosis will be seen to lie at the very heart of the origins of the animal kingdom, extending beyond insects and the diverse and fascinating marine divisions of life to the vertebrate animals, including the familiar frogs and toads, and possibly to certain aspects of humanity. There are scientists who believe we humans undergo metamorphosis in the profound and body-changing experience of puberty – and there are extrapolations of metamorphosis that may help us to understand the development of the very organ that defines our sentience, our extraordinary human brain.

For some people, such wonders are simply beyond any explanation. I understand and empathise with this feeling of awe. But I want to know how such strange and wondrous changes actually arose. This book searches for the answers to these great riddles.

# Part I

# Anomalies in the Tree of Life

*Those strange and mystical transmigrations that I have observed in silk-worms turned my philosophy into divinity ... Ruder heads stand amazed at these prodigious pieces of nature, whales, elephants, dromedaries and camels ... but in these narrow engines there is more curious mathematics. Who ... wonders not at the operation of two souls in those little bodies?*

Thomas Browne,
'Religio Medici'[1]

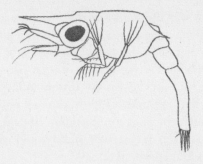

*Sponge crab larva*

# 1

# The Birth of an Idea

THE NORTHUMBERLAND COAST is as lovely as it is sparsely populated. Here, in the early 1930s, the youthful Donald Williamson would explore the lively fishing town of Seahouses, population little more than a thousand, where his father was a schoolteacher and amateur naturalist. At a time when the oceans had not yet been over-fished and polluted as we find them today, the herring boats would chug out to sea from the old harbour and return so overladen with catches that MacKay's fish shop would sell fresh herrings for a penny. To the north and south lay the great windswept beaches of Bambrugh, Beadnell, and Embleton Bay, where the lobster creels would be piled high on the harbour. On his walks, he came across bladder wracks to pop, or he collected the whorled shells of sea snails, some lined with brilliant mother-of-pearl inlays, and sea urchins and starfish, with their exotic prickly shapes and colours. But it was the sea birds that most fascinated the boy. His greatest treat, once a year, was a trip with his father to visit the bird sanctuaries in the neighbouring Farne Islands. Here he gazed spellbound at the masses of squawking gulls, ducked his head to avoid attacks by the dive-bombing terns, or was deafened by the hundreds of kittiwakes, all announcing their names with their songs in counterpoint. In these surroundings Don Williamson fell in love with nature, and with the marine aspects of nature in particular.

In 1940, at the age of eighteen, he enrolled for a degree in zoology at King's College, Newcastle upon Tyne, then part of Durham University. The Second World War had begun and his first two years there included some aspects of military training in addition to conventional lessons in zoology, botany, and chemistry. In June 1942, he enlisted in the Royal Navy as a Probationary Sub-Lieutenant, leading to training as a RADAR officer before serving in the Mediterranean theatre on *HMS Abdiel* and *HMS Antwerp*. It was an eventful experience, ending in March 1944, when he suffered a pulmonary haemorrhage en route to Taranto, Italy. He was taken to a military hospital where he was invalided out with the worrisome diagnosis of pulmonary tuberculosis, a disease that still awaited modern treatment with chemotherapy.

After brief spells in several military and naval hospitals, Williamson was dispatched far from his beloved sea, to Wooley Sanatorium, near Hexham, in Northumberland. Here, in the sanatorium library, he came across a copy of Darwin's *On the Origin of Species by Means of Natural Selection*. 'It was the first time I ever had the opportunity of reading it', he recalled.[1] Inspired now more than ever to return to his biological studies, Williamson defied medical advice and returned to Newcastle, where he completed his degree while breathing on one lung, his other lung kept out of action through a surgically induced deflation, known as an artificial pneumothorax.

Supported by a disability allowance from the Royal Navy, he went on to complete his PhD. The subject of his research was inevitably marine: a tiny crustacean known as a sand hopper, which, as its name suggests, hops about beaches in between tides.

Sand hoppers can easily be recognised from their small size and their shrimp-shaped bodies bent into a C-shape, with their long antennae continuing the curve. In Australia these include the bush fleas and leaf hoppers, which are common in gardens and undeveloped bush – "small brown critters" that leap up when disturbed. In Britain, the land-based varieties are confined to beaches, with the exception of a single species that has invaded household greenhouses.

One species of sand hopper, known as *Talitrus saltator*, particularly intrigued Williamson and would become the subject of his doctoral research. He observed how it spent the daylight hours above the high-water mark, avoiding desiccation by hiding under the sand or decaying debris and only coming out at night to feed on washed up wrack. He set out to explore the sensory cues behind their interesting behaviour. At this time biologists believed sand hoppers found their way using a kind of position sense that enabled them to detect the slope of the beach. Choosing a time when the tide was out, Williamson ferried his hoppers from the high-water mark to the low-water mark, then monitored them as they made their way back up the shore. How speedily and accurately they managed this, travelling upslope in what appeared to be straight lines! If they navigated using a sense of gradient but no sense of vision, this was an extraordinary feat. When he covered up their black multi-faceted eyes, he discovered that the sand hoppers completely lost their way. He tested this further by showing them lantern slides of simple shapes and shades and confirmed they were attracted by boundaries where dark met light. It was clear that the prevailing views were wrong. Sand hoppers navigated the beach not through sensing the gravitational pull of its slope but through visual cues.

Two years later, his PhD dissertation complete, Williamson needed to find a job. He was offered a post in Jamaica, but it would oblige him to become an entomologist. The University of Sheffield offered an alternative, but the city was land-locked and he would have to sacrifice any possibilities of working with the sea. He opted for the Isle of Man, where a post was open in marine zoology, and specifically that of planktologist with lectureship duties to Liverpool University. 'Of course, I was not a planktologist, having conducted my research on hoppers, which were semi-terrestrial. But nevertheless I applied for the job and I got it. I now had to learn about plankton'.

The word "plankton" is evocative: it derives from the ancient Greek for wanderer. And how wonderfully apt it is. Imagine a planet where ninety-nine per cent of the living space is ocean: of course, we're living on it. While we rightly assume that only two-thirds of the Earth is covered by water, any true comparison must take into account biospheric volume: the terrestrial habitat is largely two-dimensional, while the oceans are three-dimensional and, in places, miles deep. The surface waters are inhabited by tiny open-water algae, known as phytoplankton; minuscule animals, known as zooplankton; as well as a multitude of different bacterial forms that conduct their lives in a veritable zoo of unknown viruses. Plankton, in this small plant, animal, and bacterial form, are the foundation of the oceanic food web, providing food for larger creatures, such as fish and whales. Plankton also make an important contribution to life in general. Phytoplankton, such as algae, live close to the surface where, like plants on land, they capture the energy of sunlight, taking carbon dioxide out of the atmosphere and releasing oxygen back into it. In this way they play a crucial role in two of the great cycles of life, Earth's oxygen and carbon cycles, while also reducing the tendency to global warming. Others are involved in the nitrogen cycle, capturing nitrogen from the atmosphere, incorporating it into more complex organic chemical compounds, which are then fed back into the web of living interactions that create the basis of life on Earth by providing essential nutrients for all plants and animals. Zooplankton do not necessarily need light and so can live at any depth in the so-called pelagic, or upper reaches, of the oceans. They include an extraordinary diversity of forms, many of which display a spectacular if eerie beauty, best seen at low magnification and in conditions where they transilluminate light. A major component of zooplankton is the larvae of marine invertebrates, including those same sea urchins and starfish Don Williamson had collected on the Northumberland beaches.

Marine larvae are an integral stage in the metamorphoses that encompass a wide variety of marine invertebrate animals, the oceanic equivalent of the caterpillars of insect metamorphosis. In his role as planktologist, Williamson would conduct many original research studies on these larvae. But one group of marine creatures in particular would come to delight him: the crustaceans. The crustaceans constitute a major division of the animal kingdom known as a phylum, which, together with two other groups, the mandibulates (which have antennae and jaws, and includes the insects) and the chelicerates (which lack antennae and jaws) make up the superphylum of the arthropods – invertebrate animals with jointed legs.

All crustaceans have a hard outer shell and two pairs of antennae adorning their heads. They include crabs, lobsters, and shrimps as well as a bewildering variety of less familiar creatures, varying from barnacles to water fleas. When, in the 1950s, Williamson first began to study crustacean plankton, he learned that many areas remained to be explored. For example, the larvae of many of the hermit crabs found in British waters had never been described. Teaming up with a colleague, Richard Pike at the Millport Marine Biological Station on the Isle of Cumbrae, he set about filling in the gaps. In time, Williamson became a globally acknowledged expert on marine metamorphosis and, in particular, on the larvae of crustaceans. But again and again, in describing such metamorphoses, he was confronted by mysteries.

Take, for example, the life histories of sponge crabs and hermit crabs.

Sponge crabs carry living sponges above their carapaces as a means of protection, akin to the way we protect ourselves with umbrellas against the rain. The sponges provide the crabs with protection of a different sort, as a mixture of camouflage and protective shield in times of danger. If a predator threatens, the crab remains motionless under its living umbrella. If a determined predator still takes a bite, all it gets is an unappetizing mouthful of sponge, filled with prickly spicules. Sponge crabs are true crabs, slotting appropriately into the crab section of the tree of life – in scientific jargon, the evolutionary or "phylogenetic" tree. But the larvae of sponge crabs don't look like the larvae of other true crabs. Instead they closely resemble the larvae of hermit crabs, which are not true crabs but are more closely related to lobsters. It struck Williamson as peculiar that, in spite of the considerable evolutionary separation between sponge crabs and hermit crabs, their larval stages were virtually the same. He racked his brains to find a conventional explanation, but could not find one. It became something of a challenge. 'From then on', as he would subsequently recollect, 'I wanted to solve the sponge crab paradox'.

Conventional explanations of these unexpected larval similarities rested upon a principle known as convergent evolution. Consider dolphins and whales. In their streamlined shapes and the use of their tails for locomotion, these marine mammals resemble fish. This is an example of convergent evolution. Moreover, it is easily explained. These marine mammals are rapid swimmers living in the same environment as fish, the oceans, and so natural selection, which ultimately dictates the shape and movement of fish, has adapted the shape and movement of marine mammals along similar lines. However, when one contrasts and compares marine mammals and fish, the convergence is seen to be superficial. In their internal anatomies and organs, dolphins and whales bear little resemblance to fish. Like all mammals, they are warm-blooded and use lungs to breathe oxygen from the air rather than using gills to extract it from the water. In many other aspects of their internal anatomy, their separate evolutionary histories are obvious, despite the convergence of superficial body shape.

Another striking example of evolutionary convergence is the similarity between the eyes of molluscs, such as squids and octopuses, and those of vertebrates, such as fish and humans. Both are camera-type eyes in which an image is captured by a lens and focused onto a light-sensitive retinal layer. The eyes are also remarkably similar in many other aspects of their appearance and organisation. The commonalities extend to a developmental gene, known as *Pax-6*, which plays a key role in constructing the eyes of squids and fish during their embryology. *Pax-6* appears to play a similar role in eye development throughout the entire animal kingdom, from the simple eye spots of earthworms to the multifaceted eyes of butterflies to the eyes that enabled the genius of Rembrandt. This would suggest that all visually endowed animals share a distant, likely very basic, common evolutionary ancestor that first discovered a means of responding to light. But does that mean Rembrandt shares the detailed evolution of his remarkable vision with octopuses and squids? A previous generation of biologists refused to believe so. They saw the present-day similarity in the eyes of molluscs and humans as a classic example of convergent evolution. And time has proven them right, though the proof involves degrees of subtlety that were only revealed with modern tools of molecular biology coupled with the precise study of development, which led to the discovery that specific components of the eyes of vertebrates and squids develop through quite different mechanisms and from different embryonic sources. While we may well share a very distant common ancestor with molluscs, which might explain the common use of the *Pax-6* gene, the more overt similarities of our camera-type eyes are actually the result of much later convergent evolution.[2]

Convergence is thus an important, occasionally compelling, explanation of some parallel forms seen in nature. So Williamson asked, 'Was convergence a convincing explanation of the striking similarities between the larvae of the sponge crab and the hermit crab?'

Anyone who has studied marine larval forms cannot fail to be impressed by their amazing diversity of shapes and patterns. Fast swimmers, such as fish and whales, may have evolved a streamlined shape as a result of convergence through the need for locomotion in the same ecologies, but there is no discernible convergence of shape among the myriad larvae that inhabit the same streams of the oceans. This suggested to Williamson that similarity of ecological needs and constraints was unlikely to explain the similarities he was observing in larvae from two widely divergent places on the evolutionary tree. Indeed, the harder he probed the relationship between marine larvae and their place on the conventional tree of life, the more anomalies he identified.

In the late nineteenth century, scientists believed the embryonic development of any creature captured its evolutionary history, a concept first proposed by the German naturalist Ernst Haeckel as a biological law. Today developmental biologists no longer hold to this "recapitulationist" theory in an absolute sense – we now believe evolutionary adaptation can occur at any stage, including the embryo and larva – but the theory still can be useful, if treated cautiously. For example, the fertilised human egg develops into an embryo equipped with a tail and fish-like gills. Evolutionary biologists posit that a distant ancestor of mammals, including humans, was a fish-like animal that possessed gills and a tail. The human embryo could be portrayed as recapturing this stage of our evolutionary past. Many evolutionists invoke Haeckel's law to some degree in attempting to explain the changes of metamorphosis. However, Williamson saw a flaw in such evolutionary thinking when it came to the important marine phylum of prickly-skinned animals, such as sea urchins and starfish, known as echinoderms.

Part of the allure of these exotic creatures lies in their rounded and starry shapes, which, like flowers, imbue them with an aesthetic beauty so unlike most of the animals we see on land. We humans have a left and a right side. We are bilaterally symmetrical. Echinoderms don't have a left or right side; rather, like the petalled heads of the majority of flowers or like oranges, they are radially symmetrical. How strange, then, that radially symmetrical sea urchins and starfish begin their lives as larvae that are bilaterally symmetrical. This change during development, from bilateral to radial symmetry, involves one of the most spectacular metamorphoses in all of biology and it necessitates an almighty reorganisation of the larval anatomy, including skin, skeletal structures, the vascular circulation, and the structure of the nervous system. We shall look at the metamorphoses of starfish and sea urchins in more detail later, but for the moment it is only necessary to grasp the general principles.

Williamson found it difficult to imagine how a radial starfish could possibly have evolved from a bilaterally symmetrical ancestor. To his mind it seemed to imply that at some stage in its evolution the ancestor of the starfish had wiped out its inherited body plan and evolved a completely different development from scratch.

In the early days of his appointment to the Marine Lab, he had not been required to lecture to students. But from the 1960s onwards, an increasing number of biology students began to take an interest in marine biology. University students who wanted to specialise in marine biology spent a year on the Isle of Man studying at the marine station. As part of a course he developed for these students, Williamson gave a lecture on marine metamorphosis and its larval aspects – essentially the same lecture, year after year. Then, as now, he was a firm believer in Darwinian evolution. He would tell his students that, in general, most marine larvae appeared to match what would be expected from orthodox evolutionary theory, but there were anomalies that demanded explanation. Most taxonomic classifications were based on the adult forms, with the larvae assumed to complement the positions of their respective adults on the tree. The problem, as Williamson had now come to realise, was that many larval forms just did not fit in with the extrapolation of the tree of life based on the adults.

Darwin had puzzled over the mysteries of metamorphosis, and he had himself put right the mis-classification of barnacles through a study of their larvae. Indeed, in *Origin of Species* he concluded that an animal species should be seen as comprising all phases of development from the egg to the adult, each phase, including the larvae, undergoing its own separate evolutionary modifications under the influence of natural selection. The pressures of natural selection working on each of the different phases of the life cycle adapted the relevant form, whether larva or adult, to survive in its specific ecology. Or to put it another way, each phase of a complex animal life cycle, from embryo to larva and from larva to adult, would be honed by evolutionary pressures, with ultimate survival to the stage of successful reproduction – usually encompassed by the adult stage – the ultimate key to evolutionary success. In this way the free-swimming bilaterally symmetrical larva of a starfish had been honed by natural selection to fit life in the vast surface, or planktic, layers of the oceans – its central role being dispersal to as wide a geographic range as possible; meanwhile, the radial adult had been honed to fit a predatory existence on the much more geographically restricted ocean floor. But while Williamson accepted that natural selection would separately adapt larval and adult developments, he saw many anomalies in this ideal simplicity – anomalies of symmetry, of detailed development of vital organs and whole animal forms – that were too readily dismissed by his colleagues. He drew up a new tree of life for the marine invertebrate animals, based not on the adult body forms but on the various larval body forms, and he then compared his larval tree with the traditional tree based on the adults. If the larvae and adults had followed the same evolutionary trajectories – as Williamson assumed they must, following orthodox theory, which implies a linear evolutionary development, including all stages of the life form – the two trees should overlap exactly. But they did not.

'On occasions during evolutionary history', he would endeavour to explain to his students, 'larval and embryonic forms that have originally evolved in one lineage have somehow appeared in another. It is as if they jumped from one branch of the phylogenetic tree to a distinct and sometimes distant one. Of course', he felt obliged to add, 'such a thing could not have happened. The whole theory of Darwinian evolution stands firm against it. And therefore it is nonsense'.

But it was a nonsense that intrigued him year after year.

In the autumn of 1983, when he was revising the same lecture for the twenty-eighth time, he dared to ask himself a shocking question: 'What if it isn't nonsense at all? What if larvae really have moved across the evolutionary tree between species, genera and even the massively different taxonomic groups known as phyla?'

Williamson tore up his old lecture notes and drafted new ones. That year his students heard the first crude explanation of one of the most extraordinary evolutionary theories that had ever been proposed. 'I delivered the lecture in November that year and I have never had such attention to a lecture before'.

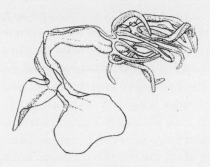

*Larva of Luidia sarsi*

## 2

# A Puzzle Wrapped in an Enigma

WILLIAMSON WAS SO INTRIGUED he set himself to solve
the puzzle. He took a long hard look at the world of marine
zoology, searching for evidence that would back up his new
idea.

Where better to look for anomalies than the familiar world
of the echinoderms? These spiny-skinned creatures constitute a
phylum all to themselves, including the starfish and sea urchins.
Adults in all of the member species have three things in common: they are radially symmetrical, based on a five-segment,
or pentaradial, plan; the various permutations of their internal
skeletons, and their stiffening spines, or external skeletons, are
all made of calcium carbonate; and they have an internal body
cavity, called a coelom, that is divided up in a distinctive manner. Now, as Williamson examined their life and evolutionary
histories in detail, he discovered a great many anomalies. For

example, he looked in detail at the life history of the starfish, *Luidia sarsi*.

This animal, which appears to have no common name, is unusually large for a starfish, growing to twenty centimetres in diameter. Ranging from sandy to salmon pink in colour, it has the typical radial body plan of the echinoderms, with five gently tapering arms. Biologists describe its skin as having a velvety texture, and they remark on a conspicuous band of white spines running along the edges of the arms. *Luidia* inhabits the muddy sediment of seashores from Norway to the Mediterranean, hunting down its prey in the dark of night. But where most starfish feed on scallops and other bivalve molluscs, *Luidia* preys on other species of starfish, a fact that is pertinent to the more discerning scallop fishermen.

*Luidia's* larva is a diaphanous sprite that grows to four centimetres long, the largest marine invertebrate larva Williamson has ever seen. The technical name for its body plan is "bipinnarian". In appearance it resembles an uprooted vegetable, with a tangle of roots at one end and two broad and fleshy leaves at the other. In fact the roots are arms covered in hairlike locomotory appendages, or cilia, which encircle the separate openings of mouth and anus, and the leaves function as the equivalent of locomotory wings. At night it feeds on algae in the surface waters, but during the day it sinks deeper, swimming through the pelagic layers with flapping undulations of its wings. Unlike the adult starfish, with its radial symmetry, the larva is bilaterally symmetrical. The adult is conceived – I cannot think of a more appropriate term – from a cluster of cells lining the internal cavity of the larva and here it grows and matures, an alien existence, independent of, and seemingly oblivious to, the larval body structures, axis, bilateral symmetry, and form, imbued with what can only be described as a complete disregard to every embodiment of its larval stage of existence. In time the juvenile starfish emerges from the still-swimming larva to settle on the ocean floor, where it begins its own independent existence hunting down other starfish for food. In the case of most other starfish the larva is sacrificed as a brutal conclusion to the metamorphosis. But in the case of *Luidia*, fate is kinder. After the adult has broken free of its tissues, the larva carries on with its independent lifestyle, swimming the pelagic waters and grazing on its vegetarian diet of algae.

This metamorphosis is so startling we are obliged to take a mental step backwards and reflect on what we have witnessed. Two separate beings, the larval and adult starfish, with radically different body forms and ecological life cycles, have developed out of a single fertilised egg. When first he read about *Luidia's* extraordinary metamorphosis, Williamson's curiosity was aroused. In his words, 'These facts needed an [evolutionary] explanation, but I thought at the time that it was up to the echinoderm specialists to provide one'.[1] No explanation had ever been proferred.

The dual development of *Luidia* takes place in a precise sequence: first we have fertilisation and the original conception, with the larva developing from the fertilised egg. Then, within the larval abdominal cavity, a new conception appears, with the adult developing from totipotent cells – the equivalent of the stem cells in humans (and other organisms) that are capable of giving rise to any body tissue. For a time, two different life forms co-exist within the one body. What, in philosophical terms, is the role of the larva? It is not quite the same as a parent – the true parents of the adult are the same as those of the larva. Perhaps it would be more accurate to visualise it as a surrogate parent. Moreover, the larval genome, present in every cell of its bilaterally symmetrical body, is identical to the genome in every cell of the radially symmetrical adult starfish. The differences seen in the two different body plans and life cycles do not derive from differences in the genes of the larval and adult cells but from the presence of two quite different developmental blueprints within the same genome. To put it another way, we need to explain how the separate evolutions of larva and adult, with their separate developmental programmes, have entered the life history of a single starfish.

Conventional Darwinian theory assumes a linear process of adaptation, through natural selection, in the evolutionary history of an organism. It's not surprising that Williamson felt obliged to ask how such conventional theory could possibly explain the utter disregard of the larval development and body plan in the adult developmental blueprint. Would a linear process of adaptation not lead us to expect that the overall body axis and symmetry of the larva should be retained in the adult? Would we not expect the development of vital organs, such as the mouth and gut, however modified, to be inherited from larva to adult? Would we not expect the same process of adaptation and modification to apply to internal organs and tissues?

Williamson doubted that conventional evolutionary theory *could* explain the metamorphosis of *Luidia sarsi*. And if it failed to explain the more bizarre and wonderful, how could it possibly offer a comprehensive explanation of the full range of marine metamorphosis?

For all the variety of terrestrial animals, the splendour of life in the oceans is breathtakingly more diverse. And where metamorphosis, in remarkable form and beauty, is a feature of two terrestrial classes, the insects and amphibians, it features, in a bewildering compass of richness and variety, in at least fifteen separate marine phyla. The mystery of marine invertebrate metamorphosis is inevitably more challenging. At the same time, the solution to this great puzzle might itself hold the key to a very great prize. Life began in the oceans. And marine invertebrates were the first animals to evolve. Since marine metamorphosis is very ancient, perhaps even primal, then understanding marine invertebrate metamorphosis might open a new window of enlightenment onto the evolutionary origins of the animal kingdom as a whole. It is little wonder, therefore, that this mystery has engaged and baffled a great many minds. We might start by asking a fundamental question of our own: why should marine invertebrate metamorphosis have evolved at all?

Five hundred million years ago, before the rise of the fish, ancient cephalopods dominated the oceans. The term "cephalopod" means a combined head and foot, but even this odd appellation belies the marvels of form and life history embraced by this class of animals. There are six hundred species of cephalopods, which belong to the most diverse of all marine phyla, the molluscs. Those ancient cephalopods are thus related to the humble snails we see in our gardens. With their jet-propelled locomotion, driven by three hearts pumping blue blood, their ability to change colour faster than a chameleon, and their highly developed eyes – as large and sensitive as any mammal – this varied class includes the chambered nautilus, the flashback cuttle fish, the oceanic squid, and the fastest and most intelligent invertebrate animal on the planet, the octopus.

Among the cephalopods, the family of octopods takes its name from the fact that its member species possess eight arms, commonly misinterpreted as tentacles. The most familiar of these are the hundred or so members of the *Octopus* genus, which inhabit the ocean floor. In these animals we find that the typical mollusc shell is either lost or internally reduced. *Enteroctopus dofleini* is the largest octopus in the world, with individuals growing up to sixty-eight kilograms. This prodigious animal inhabits the continental shelf of the North Pacific Ocean, ranging from southern California to Alaska and extending across the Aleutian Islands into Asia as far south as Japan. The mature female lays her twenty thousand to one hundred thousand eggs on the inner side of a rocky den, tending, cleaning, and aerating her brood until they hatch, between 150 to 210 days after fertilisation. And here we witness a poignant example of maternal sacrifice. This devoted mother does not eat while she cares for her young and she dies of malnutrition when the eggs have hatched. The hatchlings swim towards the surface, where they join the exotic zoo of planktic life forms.

Though often loosely called larvae, octopus hatchlings are not larvae at all. In fact octopuses, like all cephalopods, do not undergo metamorphosis. Instead they hatch as miniature adults. These minuscule octopuses inhabit the planktic layers for four to twelve weeks, drifting with oceanic currents, until they reach a mantle length of about a centimetre and a half, at which time the young settle back to the bottom to begin their adult life history, searching out a new den and growing into fearsome predators of crabs and scallops.

All authorities agree that the cephalopod molluscs inhabit the two ecologies common to most marine invertebrates – an early spell exploring the free-floating, three-dimensional ecology among the plankton of the surface waters followed by the mature life cycle in which the adult octopus must crawl and hunt in the two-dimensional ecology of the ocean floor. So how have they escaped the need for the dramatic changes of metamorphosis?

One of the most elegant ideas in embryology of the nineteenth century was undoubtedly Ernst Haeckel's recapitulationist theory. He also argued that the shapes and changes experienced by the embryo were of such critical importance to every animal species that these must predict the development of the form and structures of the same animal throughout its subsequent evolution. This was widely accepted in Darwin's lifetime, and it was extended to the shapes and changes of larval development. If the life cycles of marine invertebrates followed Haeckelian principles, there was no mystery about their metamorphosis: At some time in the distant past, the ancestor of the species resembled the larva, and this ancestor must somehow have evolved, in the linear manner proposed by Darwin, until it was transformed into the adult animal, with the remarkable life cycle we witness today. In this way, every variety of marine metamophosis would be a palimpsest of its evolutionary origins and journey.

While most modern biologists have rejected recapitulation as a universal law, sometimes the early stages of development really do give us useful information on ancestral form and structure. For instance, whales don't have legs, but they have tiny remnant bones buried in their bodies, and leg extremities begin to develop in the whale embryo before receding again. Thus if we accept that recapitulationism is far from absolute, and if we can figure out the limited circumstances in which it actually applies, the study of metamorphosis might help biologists reconstruct some important aspects of evolutionary history.

In the 1870s, Francis Maitland Balfour, brother of Prime Minister James Balfour, took on this historical challenge while studying marine larvae at the University of Cambridge. He was one of the first to contest Haeckel's theory, concluding that even the earliest embryonic stages of an animal's development were capable of evolutionary change, which would obscure or mislead any recapitulationist interpretations. He also concluded that evolution could suppress earlier developmental stages, so that the larval phase could be lost from a previously metamorphic animal life cycle. This implied that the interpretation of larval-adult relationships was trickier than ever. In an effort to explain the admittedly complex world of larval and adult relationships, Balfour made the bold assumption that metamorphosis was universally present among the earliest animals. To explain why octopuses and many other present-day animals do not undergo metamorphosis, he took the view that these must be descended from ancestors that actually did metamorphose, but had abandoned their larval stages over the eons of their evolutionary history.

Balfour decided he would call the original and most ancient of larval forms the "primary" larvae. If these had evolved in linear Darwinian fashion from the earliest animal embryos, it suggested that the forms of these primary larvae would offer tantalising evidence about the evolution of the earliest animal forms. The relevant metamorphosis would repeat its own ancestral history. But when he searched for primary larvae he encountered a problem: even if he correctly identified a larva as primary, its original form might be greatly changed as a result of evolution affecting the larva itself. He realised he would have to take his searches a major step further, comparing larval forms across a phylum, and even across the whole of the animal kingdom. Only then could he peel away the work of evolution to reveal the true ancestors of phyla, and, he dared to hope, perhaps even the primal ancestor of the entire animal kingdom.

Balfour and Haeckel were contemporaries, members of a global intellectual elite that, much as the artists, writers, and composers of their day, interacted with and listened to one another, leading at times to mutual enlightenment and at other times to furious disagreement. In the early 1870s, Haeckel had proposed, on the basis of his own theory, that the primal ancestor of all animals was a very simple marine creature, resembling the earliest stage of embryonic development – a hollow ball of cells with a primitive mouth opening into a sac-like gut. Haeckel named this the "gastraea".[2] Balfour decided he would examine embryonic and larval forms throughout the animal kingdom, looking for hard evidence of this gastraea. He would also compare present-day larval forms with living and fossil adults in different groups. Where he found common form and structure in both larva and adult, he would assume that the present adult must be closely related to the ancestral stock of the group in which the larva was found. Any such larva could then be regarded as primary.

But when he began the actual searches, the picture became increasingly uncertain. Indeed, in time Balfour arrived at the startling conclusion that most of the present-day marine larval forms, such as the larvae of the starfish and sea urchins, were not primary at all. They had somehow invaded the pre-existing life cycles of species. Extrapolating from his theory, he had to assume that in the distant past there would have been primary forms in these invaded life cycles but those primary forms had been lost and replaced by what he now termed "secondary" larvae. And while primary larvae might offer important clues to the distant ancestors of the present adult life forms, these secondary larvae, as far as he could determine, provided no ancestral link at all to the adult phyla in which they were found.

He turned his attention to the phylum of the cnidarians, which includes the sea anemones, jellyfish, hydras, and corals. These are the simplest of animals in terms of body blueprints and tissue organisation, and so it will come as no surprise to find these are also among the earliest animals to appear in the fossil record. They follow two basic body patterns: finger-like, as in the case of the hydra with its tubular shape, and medusoid, as in the jellyfish with its bell-like body and streaming tentacles. Interestingly, both these body forms are radially symmetrical in the horizontal plane. Their larvae adopt the simplest of all larval forms, a radially symmetrical solid ball of cells, the planula, which has a surface layer covered by locomotory cilia. Balfour saw the planula as a perfect example of his primary larva. If a Medusa-like radially symmetrical organism was one of the earliest animals on the scene, and if this animal had evolved from the still simpler planula, then the planula was as close as he was likely to get to the primal animal ancestor.

What did his contemporaries make of Balfour's ideas, which he published in a book in the year 1881?[3] Charles Darwin, a friend and admirer, wrote a letter to Fritz Müller, dated January 5, 1882, in which he stated:

> *Your appreciation of Balfour's book has pleased me excessively, for though I could not properly judge of it, yet it seemed to me one of the most remarkable books which have been published for some considerable time. He is quite a young man, and if he keeps his health, will do splendid work ....*[4]

However, Darwin's assessment was wrong in one regard. On February 13, 1882, the famous naturalist wrote a new letter, this time to Dr Anton Dohrn:

> *I have got one very bad piece of news to tell you, that F. Balfour is very ill at Cambridge with typhoid fever ... I hope that he is not in a very dangerous state; but the fever is severe. Good heavens, what a loss he would be to Science, and to his many loving friends!*[5]

This presaged a sad, and somewhat ironic, ending to this story. Balfour recovered from his typhoid fever and travelled to Switzerland for a period of convalescence, but, unfortunately, while attempting to climb the unconquered Aiguille Blanche slope on Mont Blanc, he died from a fall. He was just thirty-one. And as Darwin anticipated, the young biologist's death, barely a year after publication of his pioneering ideas, was a major loss to the world of science, and to the burgeoning science of embryology in particular. How poignant also that on Wednesday April 19, 1882, just two months after writing his letter of concern, Darwin himself died.

For Brian K. Hall, Professor in Biology at Dalhousie University in Nova Scotia, Balfour's conclusions, however speculative, even today provide a useful guide for embryonic study.[6] But by and large, the broader world of science soon forgot Balfour's bold work. It was a neglect that would prove costly to Williamson, when, unaware for many years of Balfour's thinking, he struggled to come to terms with similar questions and problems.

*Bipinnaria larva of a starfish*

# 3

# First Experiments

THE PORT ERIN MARINE LABORATORY on the Isle of
Man was founded in the nineteenth century by a committee of
amateur naturalists, headed by Professor William Herdman,
a zoologist from the University of Liverpool. It began with
two small buildings, but Herdman soon convinced the Manx
government that it was in their interest to expand the facility
to include a fish hatchery, a place to study marine life, and a
public aquarium. The government duly put up the money to
build a more comprehensive institution, set amidst the rocks
of a quarry on the beautiful sweep of the bay. At the time I
first visited it, in 2002, the Marine Lab had long since been
incorporated as a teaching group into the School of Biology in
Liverpool. Students came over to the lab as part of the modular
course in biology, and those with a deeper interest in marine
biology spent their honours year working with the lecturers and

postgraduates conducting various research programmes. I was surprised to discover the international nature of this research, which was often highly practical – for instance, helping colleagues to set up their own laboratory in Egypt, showing them how to manage fisheries in Chile, or assisting the development of a coral reef conservation programme in the Philippines.

Part of the remit is environmental awareness. Because of its location and length of tenure – the Irish Sea has direct continuity with the Atlantic Ocean, and measurements have been taken since 1900 – the laboratory has provided valuable data about increasing oceanic pollution and temperature. Indeed, lab measurements have helped to confirm a slight, but significant, rise in oceanic temperature of 0.5 degrees Celsius over the past century. Given the vast capacity of the Atlantic Ocean, this is important confirmation of global warming.

On a sparklingly fine day in January I found myself outside the main doors to the Marine Laboratory, listening to a little of this history and gazing across the bay of Port Erin in the company of Professor Trevor Norton, the lab's current director. Seagulls shrieked and nested in the cliffs immediately behind us. The headland on the far side of the bay is known as Bradda Head. I couldn't believe that it was blooming with colour even in the depths of winter. Around the base of a stone tower spread the golden yellows of European and Spanish gorse, the gingery rust of bracken, the purple mist of heather. 'We had two new students, one from China and one from the Philippines, and they asked me: "Who planted all these flowers?"' Norton threw back his head and roared with laughter. 'As an Irishman you'll understand – there's magic in it.'

Magic captured my enchantment exactly.

Norton combines a love of science with the art of writing. He has penned three delightful volumes around his memories of marine biology. These include *Reflections on a Summer Sea*, which is based on his youthful experiences among a menagerie of dedicated ecologists working through a series of summers around Lough Ine in the southwest of Ireland. In reading the book, I was amused though hardly surprised to discover that the English visitors to Lough Ine were even more eccentric than the native Irish. 'In fact the scenery here,' he informed me as we faced the decorous headland, 'is very like that of Ireland. It is full of these lost beaches and beautiful forgotten valleys.'

When he first formulated his hypothesis of larval transfer, Williamson realised it was not enough merely to discover anomalies that might be explained by his hypothesis: he needed to find a plausible explanation for how a larval form might be transferred from one branch to another on the tree of life. Of course, he was not thinking that a larva had literally crawled or swum across the evolutionary tree. The transfer he had in mind was not that of a whole organism but of the genetic blueprint for that larval shape and life cycle, which would need to be carried, through an evolutionary mechanism, from one life form to another. He considered various options before concluding that it must occur through the biological phenomenon known as hybridisation, which involves the sexual union of parents from different species. Hybridisation, in bringing two different genomes into the genetic inheritance of a single offspring, offered the only plausible evolutionary mechanism for his theory of larval transfer.

One of the beautiful if unfrequented beauty spots referred to by Professor Norton is Scarlett Point, a projection of grey carboniferous limestone ledges broken through by vertical stacks of dark volcanic lava. A curved grassy path leads to a series of brackish pools out of reach of the tides on calm days but drenched in spray when the weather plays rough. These pools are inhabited by an unusual shrimp, *Gammarus duebeni*. This was the animal Williamson chose to study in his first hybridisation experiment.

*Gammarus* means "lobster-like". If this creature has a biological claim to fame, it is the fact that it is the only British species of shrimp to breed in a wide variety of salinities, from freshwater to highly salty water, so that it is commonly found in pools subject to evaporation around the high-water splash zone. The male *Gammarus* is larger than the female and comes equipped with specially adapted "toothed" feet. He uses these to grip the carapace of the female when they mate, so she dangles in a love-hug below him. They remain locked in this intimate way for days. After mating, the female moults and lays eggs into her brood-pouch, which cues the male to squirt his sperm into the same chamber. In many cases, the female will not lay her eggs unless she is being carried by a male. Later, when I questioned him, Williamson was a little vague in recounting why he chose this particular shrimp to conduct his first experiment. 'I can't remember how I got the information – whether I did it by trial and error – although *Gammarus dubeni* will usually lay eggs when it is coupled with the male, it will also lay eggs even when isolated. Perhaps I was influenced by the fact it is extremely hardy, thus likely to survive the interference of prolonged biological experiment. And I knew where to get some'.

In 1986, equipped with his daughters' shrimp nets, he made his way to the pools at Scarlett Point, where he fished for the shrimps. They were not easy to see in the dark-reflecting water, camouflaged a nondescript greyish brown and, even as full-grown adults, no more than two centimetres from antennae to the tips of their tails. Amphipod crustaceans such as *Gammarus* do not undergo metamorphosis, developing from egg to adult without undergoing a larval stage. This was critical to what he had in mind: an experiment that had never been performed before in the history of marine biology. Williamson wanted to see what happened when he mated this marine shrimp, which had no larva, with a sea urchin that metamorphosed through a larval stage. 'So I set up a tank in my room at the marine biology laboratory, with *Gammarus dubeni*.'

He was pleased to find the shrimps bred in the tank without any problem. Soon he had hundreds of *Gammarus* ready to begin his experiment. He began to isolate those females that seemed ready for egg laying. This he could detect because they would, first of all, moult. At this time the laboratory was equipped with a public aquarium, which contained plenty of sea urchins. Williamson fished out two or three of the common British species *Echinus esculentis*. Globular in shape and a bright pastel pink in colour, this attractive animal grows to fifteen centimetres on rocky shores, where it grazes on algae and encrusting animals. He didn't know how to distinguish male from female among the sea urchins, but he knew that if you inverted ripe specimens over a beaker they would reveal their gender by laying either sperm or eggs. This he now accomplished in large quantities without harming the creatures. Soon, he had what he wanted, a copious supply of sea urchin sperm. He placed female shrimps into a concentrated suspension of sea urchin sperm and waited for them to lay their eggs.

It took no more than a day or two for the female shrimps to start laying. But now he was faced with the problem of getting hold of the fertilised eggs, which were hidden away in the brood pouch under the thorax, to see if they would yield developing embryos. Using two needles, he pried open these private quarters and teased out the eggs without damaging the shrimps.

Animals and plants are separated into distinct kingdoms in the taxonomic classification of life. Phyla are the next great divisions below those of the kingdoms. All the members of a phylum have some defining characteristic in common, often a characteristic that is very basic indeed. For example, we humans belong to the phylum Craniata, from the Greek *kranion*, which indicates that all the phylum members come equipped with a skull. The craniates, together with the urochordates and cephalochordates, make up the super phylum of chordates, all of which possess a nerve cord within a flexible spinal column. Within the craniates, we humans belong to the class of mammals, which contributes just 2500 species to the 45,000 species of craniates, the latter including fish, amphibians, reptiles (including dinosaurs), and birds. The differences between phyla were laid down more than five hundred million years ago, back in the evolutionary period known as the Cambrian, when the major branches of the animal tree evolved. In other words, members of different phyla are separated from each other by enormous distances in terms of evolutionary time, as well as by major differences in their physical and genetic make-up.

Williamson's shrimps, *Gammarus dubeni*, belonged to the phylum of the crustaceans while the sea urchins, *Echinus esculentis*, belonged to the phylum of the echinoderms. A single observation will bring home the extent of the differences between the parental species that Williamson was bringing together in his hybrid experiment. *Gammarus dubeni*, like all of the members of the phylum Crustaceae, has a left and right side; it is bilaterally symmetrical. Sea urchins, like all of the members of the phylum Echinodermata, are radially symmetrical. Hybridisation usually involves crosses between closely related species. In conventional biological thinking, there should be no offspring from such a disparate sexual union.

Each morning, and several times during the day, Williamson would pore over the eggs with the aid of a microscope. Many survived for no more than a day or two, but a minority survived, for periods ranging from a few days to one week. An even smaller minority survived for more than a week. Survival was not enough; he was looking for evidence of development. The outer membrane of shrimp eggs is a good deal less transparent than the eggs of echinoderms and, however carefully he scrutinised the eggs, he could not distinguish cell divisions in the embryos. Still, after a time he could see a spherical form emerge in two of the eggs. He could also make out that the embryo was spinning rapidly, in a way that suggested it was being propelled with cilia. Crustaceans like *Gammarus dubeni* do not possess cilia, at embryonic or any other stage. This means that a normal shrimp embryo should not be mobile. The movement of this tiny unknown offspring had to represent some unusual development.

He waited to see if either of the two eggs would hatch out, so he could inspect the resulting free-swimming life form. But neither hatched. He attempted to dissect off the egg membrane without any success. When he tried to release the spinning form for dissection, whatever was inside disintegrated as soon as the egg membrane was ruptured, leaving him with an unrecognisable pulp. He considered the possibility of photographing the larvae inside the eggs. He had access to photomicrographic microscopes but the movement within the eggs was so rapid, all he could document was a blur. Cinefilm photography down the microscope was not available to him, so he could not record it. 'It was frustrating but at the same time the results of this single experiment gave me great encouragement. I thought I really am on the right lines and hybridisation is the answer. This must be the way in which some animals acquired what appear to be the wrong larvae.'

People commonly assume that new scientific theories arise in the minds of their inventors much as we come across them in lectures, articles, and textbooks, long after the theory was first conceived. In reality, nothing could be further from the truth. New scientific ideas, much as original creativity in painting or writing, are more likely to arise as inchoate ponderings, triggered by some flash of inspiration or some unlikely observation. This is exactly what we see with Williamson, where a new line of thinking was triggered by anomalous findings in a routine field of work. Scientists also tend to be cautious in their reactions. How many potential breakthroughs never come to fruition because their originators lose their nerve at the first hurdle? The temptation will be to put their new ideas aside for a while – witness how Darwin pondered over his evolutionary theory for decades before he published it. In the early stages scientists are likely to search around to see if, maybe, somebody else has got there before them. If they encounter silence, they may gain confidence that their idea truly is original and decide to take things further. They become sufficiently brave to test their ideas and sometimes this bears fruit. Through such predictably human mixtures of inspiration, fearful uncertainty, and thrill of exploration, creative ideas eventually emerge into the light of day.

Now, however frustrating the shrimp-urchin experiment, Williamson had a better grasp of the difficulties he faced. Before setting up any new experiment, he would have to sit down and think even harder about his methodology. He needed to pose his question in such a way that whatever result he obtained gave him a clear-cut answer. To take it still further he would need to overcome the limitations of the facilities then available to him at the Marine Laboratory.

Critically, the Marine Lab could offer him little or no molecular biological backup. Although he was well aware of the need for genetic confirmation of his work, Williamson could accommodate no subtleties of chemical or molecular analysis here on the Isle of Man. He had no option but to pitch his initial experiments at a very basic level, not much different to the type of experiments performed by Darwin or Huxley during the great Victorian era of observational biology. And unlike these illustrious predecessors, he comforted himself with the thought that if he could but prove he was on the right track, he might then enlist the aid of other scientists in other centres equipped to complete the essential genetic analyses.

If the Marine Lab had its weaknesses, it also had its strengths. Immediately to hand were huge briny tanks from which scallops, lobsters, and various types of flatfish and round fish were routinely harvested. For a hundred years biologists working here had taken a special interest in the production and study of larvae, some of which are very difficult to rear in captivity. This practical foundation, together with his own extensive personal experience of larval biology, was the logical place for Williamson to turn. After giving it some thought, he turned his focus to an animal that had long intrigued the world of biology – the humble sea squirt.

Sea squirts are regarded as distant cousins of all vertebrate animals, though you might be forgiven for not seeing the resemblance at first sight. They have no heads, eyes, or even limbs. They are often portrayed as hollow flasks, or worse still, mere bags of flesh, squatting unadventurously on the marine substratum, yet for the biologists who study these "ascidians" this belies a subtle beauty and complexity of form and development. I well remember my own sense of surprise at first observing the common sea squirt, *Ciona intestinalis*, illuminated by light. I was enthralled by what appeared for all the world to be the most exquisite narrow-bodied teapot, walled of a gossamer-thin semi-lucent crystal and gilded around its delicately fluted rim and spout. It took a second, and then maybe a third, hard stare before I could accept that I was looking at a living creature. Sea squirts are members of the phylum Tunicata or Urochordata – the latter, Greek for "chordate with a tail" – and the adults metamorphose from a tadpole larva. It is in this remarkable larval creature, which ranges in size from a mere visible speck to one centimetre, that we find the tell-tale link to the vertebrates. However much they resemble the tadpoles of frogs and toads, the larvae of sea squirts are not directly related to amphibians. Nevertheless, sea squirt tadpoles do have the equivalent of a very primitive brain, a nerve cord, and a heart, and, like amphibian tadpoles, they swim in an elegant fashion via undulations of their muscular tails.

Adult sea squirts grow up to thirty centimetres in length, some living singly while others form colonies that swarm over hard surfaces, such as tropical reefs, rocky bottoms, and ship wrecks. As a snorkelling swimmer, you might glimpse the mouth – the teapot rim – opening and closing. Take a closer look and you might spot a clear reminder of its vertebrate linkages: the flash within of eight white teeth. Cilia around the rim set up a current that sucks a feast of phytoplankton into the stomach; the waste is squirted back out through the spout-like second opening.

Given the adult form and lifestyle, it will come as no surprise that before biologists ever linked them to their tadpole larvae, sea squirts were wrongly classified – as molluscs and even as marine worms. Only when, in 1866, the Russian biologist Alexander Kovalevsky recognised that this "gilt-rimmed teapot" developed through a tadpole larva did biologists realise they were looking at a chordate. For Darwin, it was a discovery to be celebrated, for 'it seems that we have at last gained a clue to the source whence the Vertebrata were derived'.[1] Kovalevsky was also the first to observe the dramatic, if brutal, metamorphosis, in which the more obvious chordate structures of the tadpole are sacrificed during the transformation to the bottom-fixated adult in a powerful melodrama enacted, in the depths of the oceans, on a Lilliputian stage. There are reasons for thinking that the present larva is little more than a pastiche of its former existence. So impoverished has the tiny creature become, mouthless and minuscule, stripped of all the trappings of any independent existence, even stripped of the ability to feed for itself, it resembles a wraith going about its slave-like duty.

Unlike the starfish, with its extraordinary metamorphosis from bilateral to radial symmetry, the sea squirt is bilaterally symmetrical throughout its life cycle. Logically, we should expect the adult sea squirt would develop in a simple linear fashion from the basic scaffold of the larval structures, with the adult body plan following the larval axis, and growing, however extravagantly, out of the tissues and organs of the larva. And, indeed, at first all seems to be going to plan. Inside the fertilised egg, the embryo develops to the primal stage of the simple ball of cells. But from this point onwards, in the words of Professor E.J.W. Barrington, who studied the metamorphosis with undisguised astonishment, 'two independent developmental mechanisms are now operating side by side, the development of the larval structures proceeding virtually independently of that of the permanent ascidian organisation'.[2] Ignoring the larval body shape, and independent even of its essential long body axis, the adult muscles its way to life from inside the larval tissues, like an alien invader, a parasite devoted to its own selfish ends.

To begin with, the tadpole is a little ahead, but the adult soon catches up. Even the larval nervous system, including the simple brain – which might be expected to develop into the adult nervous system – is shrugged aside; the two separate brains and nervous systems coexisting, side by side, or one above the other, all in one tiny body. Soon – the period may last anywhere from a few hours to a few days – the tadpole abandons all pretence of an independent existence and heads for the very different ecology of the ocean floor. Here it discovers a suitable flat surface and glues its head to it, in anticipation of the forthcoming sacrifice. Within minutes, the beautiful tail is destroyed from within, the skin separating away from the internal tissues, its fins jettisoned. The primitive notochord – a poignant link to the chordates – is torn apart, the covering sheath ruptured, and the contents gobbled up by phagocytes. Even the tadpole brain is attacked and devoured, though its annihilation takes longer, days rather than minutes. Meanwhile the skin cells of the tail are transformed and recycled until all that remains is the emergent form of the adult sea squirt, ready to begin its sedentary existence on the ocean floor.

Like *Luidia sarsi*, this extraordinary development prompts important philosophical questions. This is very different from, say, the metamorphosis of the tadpole to a frog, where the frog develops out of the basic form of the tadpole, however much is sacrificed. The adult sea squirt completely ignores the body structures and development of the tadpole larva. How, in such curious circumstances, do we define the meaning of individual life – or death? Should we regard the tadpole as a life form in its own right, which is ultimately sacrificed? Other species, known as larvaceans, which are included in the same phylum as the sea squirts, exist as free-swimming planktic tadpoles in their adult forms. However much we attempt to rationalise it from a philosophical standpoint, the implications jar – they seem alien to our deep-felt human instincts. Judging from our human experience, and sensitivities, this thrilling yet disturbing cycle of birth, sacrifice, and rebirth challenges our essential concept of being. It is little wonder that this particular metamorphosis remains today, as it has been for more than a century, the focus of enormous scientific scrutiny. For this reason, the sea squirt genome was one of the first to be unraveled, its developmental pathways dissected and spread over thousands of computer screens, as twenty-first century science explores how this humble tadpole larva, presumed by many biologists to be close to the hypothetical ancestor of all chordates, came, in the words of Ian Meinertzhagen and Yasushi Okamura, to be the bearer of our very own 'chordate brain in miniature'.[3]

Sea urchins, with their marvellous, if equally brutal, metamorphosis, are no less challenging than sea squirts. Urchin eggs hatch into the planktic waters as ciliated balls of cells, which, in turn, develop into tiny prism-shaped bodies before extruding eight arms that are internally stiffened with a rod-shaped skeleton and covered in locomotive cilia. These are known as "*pluteus*" larvae, so-called because the German naturalist who first described them, Johannes Müller – who was perhaps less familiar with the feathered shuttlecock – thought they resembled an artist's easel. The adult sea urchin also develops, alien fashion, from pluripotent stem cells in the abdominal cavity of the pluteus. As with the starfish and the sea squirt, the adult and larval sea urchin are different life forms, with their own body shapes, means of locomotion, and methods of feeding, and there is a programmed sacrifice of the larva for the benefit of the adult. For Williamson, the marked differences between the two great marine metamorphoses – the sea squirt, with its tailed tadpole larva, and the sea urchin, with its pluteus larva – set his mind on a new and challenging experiment, one that offered an extreme test of his theory. He would attempt to fertilise a sea squirt, from the phylum of the urochordates, with a sea urchin, from the phylum of the echinoderms.

Plentiful supplies of the sea squirt *Ascidia mentula* were being cultured in one of the seawater storage tanks at the Marine Laboratory. This species, shaped like a fat sausage some eleven centimetres long, would provide the eggs. In 1985 Williamson enlisted divers, wearing aqualungs, to enter the tank, where they scraped the sea squirts off the walls and floor, providing him with his maternal species. For the paternal species, he chose the familiar urchin *Echinus esculentus*, enlisting the same divers to wade into the bay of Port Erin, where they collected dozens of animals from a ruined breakwater just in front of the lab.

As before, he milked the sea squirts of their eggs by holding them over a bowl and squeezing their bodies. There was, however, a complication: sea squirts are hermaphrodite. If he was unlucky, he collected a cloud of sperm; if lucky, hundreds of eggs. Sometimes he would get both in his collecting bowl, though thankfully this was rare. The eggs were visible to the naked eye as white spots, and he readily harvested them with the aid of a magnifying glass. The hermaphrodite sea squirt was also capable of self-fertilisation, so he had to observe the eggs over several hours. If they started to divide – the first evidence of prior fertilisation – he discarded them. With a nylon mesh, he filtered out those eggs that did not develop, ending up with about two hundred. He put these into an observation dish, poured urchin sperm over them, and waited to see what would happen. As he subsequently recalled to me: 'It was beyond my wildest dreams that I would ever get a *pluteus* larva but I had hopes, at the very least, of getting a ciliated larva.'

The experiment did not work the first time. But after four or five attempts at cross-fertilisation, some of the hybridised eggs started dividing. He could make this out under a low power microscope, needing no more than ten times magnification. In most cases the eggs divided a few times and then stopped. This was disappointing, but it was enough to encourage him to try again.

Two years passed as he laboured at a series of cross-phyletic hybridisation experiments. He grew better through practice, in effect serving his hybridisation apprenticeship. By 1987, one of the experiments resulted in two eggs hatching out into spherical larvae. This was encouraging, since sea squirt eggs should normally hatch as tadpole larvae. These hybrid hatchlings looked more like the hatchlings of sea urchin eggs, which emerge as radially symmetrical, hollow, ciliated spheres. He watched the larvae develop, monitoring how they lost their radial symmetry to become prism-shaped, with its implication of bilateral symmetry, and then further, when they sprouted arms. This was thrilling to observe, since it was exactly what he would have anticipated during the pluteal larval development

of a sea urchin. But these larvae had come from the eggs of a
sea squirt. Williamson was convinced that he had transferred
the pluteus larva of a sea urchin into the metamorphic life cycle
of a sea squirt – he had moved a developmental programme
from one phylum to another. Standard practice demanded that
he repeat the experiment.

The sea squirt normally breeds throughout the year, but the
urchin only breeds in late winter and early spring. His next
test would be postponed to the next available urchin-breeding
season. In 1988 conditions proved right and he repeated the
experiment. He was delighted to witness exactly the same outcome. 'The larvae were very like echinus larvae but developed
from ascidian eggs.'

The following year, he conducted another experiment, with
much the same result. In all these trials, the larvae developed
to the pluteus stage but failed to develop any further. By then
they were about a month old. For a cross-phyletic hybridisation
experiment, it was a remarkable achievement. A few years earlier, Williamson had first proposed that larval transfer might have
taken place in nature through hybridisation between different
species. At the very least his experiments appeared to show that
this was possible. Anyone who has watched marine wildlife programmes on television will have been impressed by the hit-and-
miss nature of "broadcast spawning", when vast quantities of
eggs laid by the females take their chances with clouds of sperm
from the males, as they float and drift in the surface waters. In
such circumstances, the sperm of one species would have ample
opportunity to come into contact with the eggs of another. He
had been successful in just a few hugely optimistic experiments.
What opportunities nature would have had, over the vast time
periods of evolution, for countless such experiments! His theory
required only an occasional success with reproductively viable
offspring to bring together the genetic programming of two
different species into a single life cycle.

For more than a hundred years, all explanations of marine metamorphosis, no matter how varied, had played, like musical variations, about the same elementary motif. Evolution worked slowly, along the straight lines of linear Darwinian descent, with natural selection operating on changes based on mutation – which itself presupposed that change arose strictly within individuals through copying errors in the genes when cells divided. Now, if they heeded Williamson, biologists were offered the opportunity of thought and debate at a more fundamental level – the potential for relatively sudden change, change that came from the sexual merging of pre-evolved and very different genomes. Such a broad perspective should stimulate science. But Williamson did not kid himself that the conservative world of evolutionary biology would welcome a radically different interpretation of common assumptions. Nevertheless, he felt ready to publish his theory, and the results of his experiments, in the hope they would be considered without prejudice.

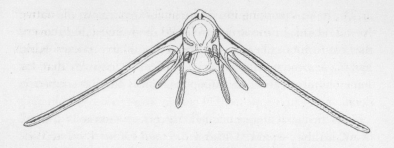

*Pluteus larva of a brittle star*

4

# The Price of Iconoclasm

IN JUNE 1985, Williamson wrote out his "larval transfer theory" in the form of a scientific paper. The title was "Incongruous Larvae and the Origin of Some Invertebrate Life-Histories". When he submitted it to the *Biological Journal of the Linnean Society,* one reviewer favoured publication while another opposed it. The editor refused full publication but offered to print a short note, summarising Williamson's ideas. Williamson declined. In October he submitted the same paper to the *Journal of Natural History*, and once again one referee was in favour; two were against publication.

In the midst of these rejections, Williamson came across an intriguing article that the marine biologist H. Barraclough "Barry" Fell had written many years before.[1] An internationally acknowledged expert on the echinoderms, Fell then held a chair at Harvard; but in 1941, when he had written the article, he was working in a more junior capacity in his native New Zealand. In the article Fell told of an unusual discovery that his former colleague and professor, Harry Borrer Kirk, had made some twenty-five years earlier – a discovery that, for some obscure reason, had never previously been described in detail.

Kirk regularly perambulated the rocky coastline of a small headland that separates Island Bay and Ohiro Bay, at Wellington, on the northern shore of New Zealand's Cook Straits, a site well known for southerly gales. One day in 1916, during a period of extremely low spring tides, he came across clusters of fertilised eggs attached to some stones. Kirk had no idea at the time what marine creature had laid them, but he brought them back to his laboratory and monitored their development. The original eggs proved to be no isolated find. On a number of occasions over the ensuing twenty-two years, he and his colleague Fell discovered similar clusters in the same spot, thereby establishing that the annual spawning season for the mystery species was in the latter half of August, the eggs being laid during the high tide but only becoming visible at low tide.

That first year, 1916, Kirk recorded his initial observations: 'The eggs are spherical, 0.5 mm in diameter, each with a perfectly transparent, thin, but extremely tough chitinous envelope'. So tough was this covering that Kirk couldn't remove it without crushing the embryos. And in spite of the envelope's transparency, he could not discern what was really going on inside the embryo because the embryo itself was opaque. His only recourse was to sacrifice some of the embryos and dissect them – but this would have to keep until the following year. In the meantime, he was content to watch and record whatever developed.

To his astonishment, the eggs hatched out not as larvae but as junior adults, a development known as oviparity. More astonishing still, the junior adults matured to a previously unknown species of brittle star, a separate class of echinoderm closely related to starfish, and easily recognisable because of their elongated whip-like arms, which they use for locomotion over the ocean bottom.[2] Kirk wrote up his discovery in a preliminary paper in which he recorded the measurements of the eggs and the details of the hatching process, together with some brief notes on this puzzling species.[3] At first he thought he had found an unusual example of the Australian brittle star, *Ophionereis schayeri*, but *Ophionereis* metamorphoses through a non-feeding larva, whereas this specimen had no larva. Over time he realised that the development of his mysterious brittle star, from hatchling to adult, was 'absolutely direct', with 'no evidence whatsoever' of a metamorphosis through a larval stage. As far as I am aware, the species of brittle star is still unattributed, but has come to be called Kirk's brittle star or, as Williamson labels it, "Kirk's ophiuromorph".

When Fell, in his turn, reared specimens from eggs in small dishes through which seawater was slowly circulated, he confirmed, in finer detail, that the development was markedly different from what he would have anticipated in a brittle star. To understand what he was observing, indeed to grasp its importance, we need to know a little more about the development of the echinoderms.

Until recently, the phylum of the echinoderms consisted of five classes: the sea urchins, the starfish (marine biologists prefer to call them "sea stars", since they are not fish), the brittle stars, the combined feather stars and sea lilies, and the sea cucumbers. Recently a sixth class, known as the sea daisies, has been added – though some biologists think they should be included with the starfish. Sea daisies have no larvae, and they are radially symmetrical throughout their development. Of the species in the five remaining classes of echinoderms that do undergo metamorphosis, the adult forms are always radially symmetrical, yet the larvae are always bilaterally symmetrical. For biologists who assume a linear evolutionary history, this dramatic change of symmetry within a single life cycle poses a dilemma. Given that the internal organisation, and thus the genetically programmed development, of a radial creature is very different from that of a bilaterally symmetrical creature, how, as part of a linear evolution, did the genetic blueprint and structural organisation of a bilaterally symmetrical larva evolve into that of a radially symmetrical adult?

The simple truth is that there was no generally accepted explanation for this dilemma at the time Williamson was formulating his hypothesis of larval transfer. It was assumed, based on larval symmetry, that the echinoderms had evolved from simpler animals with bilateral symmetry. Somehow – and it was further assumed this would have come about through mutation and adaptation via natural selection – the phylum had evolved a radically different radial symmetry. Because the echinoderms have an upper and lower surface, this radial symmetry is not completely spherical; it is radial in the horizontal plane. Or, to be more precise, it is pentaradial – divided in the horizontal plane into segments, rather like an orange, based on a rule of five, so there are five, or multiples of five, segments.

How then, given the lack of orthodox explanation, can we attempt to examine this dilemma further? One avenue we might reasonably explore is the mechanics of the metamorphosis. And, to put it mildly, these are bizarre.

We have already seen, in the example of the starfish *Luidia sarsi*, how in echinoderms with larval development, the juvenile adult begins as a cluster of cells within the larval body. We witness two distinct developmental programmes arising from the same original fertilised egg and taking place within the tissues of a single life form. The metamorphosis is accompanied by massive internal change coupled with catastrophic destruction of the larval tissues. Huge chunks of the larval body, its tissues and organs, are digested away and reabsorbed, or simply discarded. There are similarities in this "catastrophic" form of metamorphosis to what is seen in the pupal stages of certain orders of insects. Williamson asked, not unreasonably, how such dramatic changes in symmetry and such dramatic reorganisation, involving wholesale destruction and waste, could have evolved through any linear process of evolution.

To dig deeper into this riddle, and to understand it in light of Fell's observations of Kirk's brittle star, we need to know a little about embryonic development, and how it links to the basic branches of the tree of life.

The fertilised egg is spherical, especially so when supported by the buoyancy of oceanic water. Thus animal life begins with complete radial symmetry. Fertilisation triggers the process of cell division within the egg, leading to the formation of a hollow ball of cells, known as a blastula. Echinoderms usually hatch from the egg into the ocean as blastulae, their minuscule bodies ciliated to allow locomotion. In the next step of development an organised sequence of cell movements within the blastula leads to the formation of three germ layers. The outer layer, or ectoderm, will form the skin and nervous tissue; the inner layer, or endoderm, will form the lining of the gut and internal organs; and the cells in the intermediate layer, or mesoderm, will give rise to internal tissues and organs such as muscle, bone, and heart. A dimple now appears on the surface of the blastula and cells tuck into it to create a tunnel into the interior. The dimple erupts into a blastopore, which opens onto a tunnel that itself develops into a primitive gut, marking the transformation of the cells from blastula to gastrula.

What happens next sets the stage for the divergence of the two fundamental trunks or pathways of animal evolution at the base of the animal tree of life. For those member species of one great trunk, henceforth known as protostomes (from "proto" for first and "stoma" for mouth), the blastopore develops into the mouth. This protostomal lineage gave rise to the present-day molluscs, marine worms, insects, spiders, and crustaceans, such as crabs and lobsters. For the member species of the other great trunk, the blastopore develops into the anus. In these species, which includes our own, the tube-like gut breaks through to make a second opening, and it is this second opening, and not the blastopore, that develops into the mouth. These species are known as deuterostomes (from "deutero" for second and "stoma" for mouth), and they include the chordates, and thus all mammals.

These two lineages, which date back to the earliest origins of the animal kingdom, embrace a puzzle when it comes to the echinoderms. Whereas the great majority of the marine invertebrates belong to the protostomes, those most colourful and unusual of the marine invertebrates, the echinoderms, are classed as deuterostomes. This placing of the echinoderms in the deuterostome trunk of the tree of life is entirely due to their larval development. In this respect, Kirk's brittle star is anomalous.

Just like our own human, vertebrate embryo, the embryo of an echinoderm develops as a deuterostome. But when Fell investigated Kirk's brittle star he observed that its development did not follow a deuterostome pattern. The fertilised egg developed to a radially symmetrical blastula and this, quite normally, developed to a radially symmetrical gastrula. The initial tunnel leading from the blastopore was unusually small, and the primitive gut that extended from it soon filled up with cells. The blastopore itself, though its position could be seen as a dimple on the surface, now led nowhere. After six or seven days, five grooves radiated out from this blind dimple and the embryo became radial, with five segments separated by grooves running towards the midpoints of the sides. This species of echinoderm appeared to be radially symmetrical throughout its life cycle.

Over the next ten days, however, Fell observed a novel development. The adult brittle star formed directly from this pentaradial embryo. While this was occurring, the blastopore re-established a tunnel to the interior, which developed into the stomach and intestine. There was no second opening: no deuterostomal mouth. In Fell's words, 'The blastopore occupies the position of the future mouth. This is, of course, contrary to the usual fate of the blastopore in [brittle stars] since it normally gives rise to the anus of the larva'.

Fell arrived at two possible explanations for what he had seen. One was that the ancestral stock of this strange new brittle star had always developed directly. But if this was the case, direct development in this species of brittle star was pentaradial and protostomal. The other possibility was that his brittle star had evolved from an ancestral species that metamorphosed in the normal manner of brittle stars, but subsequently the larval stage had been suppressed. Larvae were known to be suppressed by cold climates and he wondered if this putative ancestor had lost its larva in response to the severity of the most recent Ice Age. There appeared to be a hint of larval suppression in the vestigial tunnel leading from the blastopore in the early gastrula, but if Kirk's brittle star really did evolve from an ancestor that metamorphosed, the suppression of the larval phase was unusually complete. Further, there was no escaping the fact that the blastopore became the mouth.

Half a century later, Williamson rejected any explanation based on larval suppression, pointing to Fell's own findings as evidence. Echinoderms with larvae should develop along deuterostome lines, with a bilaterally symmetrical larva metamorphosing to a pentaradial adult. This brittle star, without a larva, seemed to be developing in an exclusively pentaradial manner along classical protostomal lines. To all intents, Kirk's brittle star was a protostome, which would place it on the opposite branch of the evolutionary divide from its deuterostome kin.

An additional observation adds weight to this interpretation. Another important difference in development distinguishes deuterostomes from protostomes. In deuterostomes, the coelom – what we call the abdominal or peritoneal cavity in humans – usually forms by budding off from the ancient gut. Imagine a blind sac budding out from the gut wall, ballooning inwards and gradually expanding and swelling, until it fills the internal space, meanwhile folding over the internal organs to form a covering layer. In protostomes, the coelom doesn't grow out of the gut but forms out of a split in the tissue known as mesoderm. This pattern is known as schizocoely, from the Greek schizo for split and coelom for body cavity. In Kirk's brittle star, the body cavity develops from a split in the mesoderm rather than from a budding of the gut wall. This meant Kirk's brittle star was not merely a protostome but a schizocoelous protostome – a point that might seem obscure, but is of fundamental importance to evolutionary and developmental biologists. The division of the animal kingdom into deuterostomes, which develop through gut-budding, and proterostomes, which develop through meso-derm-splitting, is the very basis of the primal bifurcation of the tree of life.

How then had Williamson's more orthodox colleagues come to terms with Fell's incongruous discoveries? When Williamson studied the literature he confirmed that biologists had indeed been startled by this curious report of brittle star development. Inevitably, many, such as Libbie H. Hyman, had favoured Fell's second hypothesis, believing Kirk's new species must represent an extreme case of suppression of the larval phase.[4] But this ignored the schizocoelous nature of its development. Moreover, where larval suppression was a familiar finding with some echinoderms, in no other case did this lead to the com-plete abandonment of the deuterostome lineage. For both these reasons, Williamson deemed larval suppression to be an implausible explanation. In Fell's paper, Williamson also came across another example of direct development in a brittle star. As long ago as 1891, an Italian biologist, A. Russo, had described a protostomal development in another brittle star, *Amphipholis squamata*.[5] Russo's findings had been dismissed as 'improbable in the highest degree'.[6] In 1968, some twenty-seven years after his original paper, Fell noted that brittle stars of two other families probably developed in a similar manner to Kirk's species.[7] But nobody appeared to have followed up Fell's observation. Williamson subsequently discovered two more instances of the same anomaly: The eggs of the sea daisy, *Xyloplax medusiformis*, develop directly within the ovary and never show bilaterally symmetrical features, although the fate of the blastopore, whether anal or oral, is currently unknown;[8] and *Abatus cordatus*, a sub-Antarctic heart urchin, broods its eggs, giving birth to live young, and develops as a protostome and schizocoel.[9]

To Williamson's thinking, Kirk's brittle star was the exception that proved the rule. According to his theory, the echinoderms had originally developed to full adulthood as radially symmetrical animals without ever going through a bilaterally symmetrical larval phase. But before discovering Fell's paper, he had assumed that all such directly developing echinoderms were long extinct. Now it seemed he had been mistaken. And this in turn reinforced his view that the evolutionary origins of the phylum of the echinoderms were different from what was commonly assumed.

Adult sea urchins bore scant resemblance to adult brittle stars. Yet both classes metamorphosed through similar pluteus larvae. The echinoderms that metamorphosed through larvae were classed as deuterostomes, but, as he now proposed, these anomalous brittle stars and heart urchins, which developed without metamorphosis, should really be classed as protostomes. How could this be reconciled with orthodox theory, which proposed that both classes of echinoderms shared a common ancestor from which they inherited their developmental pathways? In analysing his own findings, Fell could hardly miss those same anomalies. Indeed, he concluded that any close relationship between sea urchins and brittle stars, as inferred by larval features, was now too preposterous to warrant serious consideration.

It seemed to Williamson that his larval transfer theory might explain the long-standing enigma of how a bilaterally symmetrical larva metamorphoses to a pentaradial adult. Where most textbooks of invertebrate zoology proposed that echinoderms evolved from bilaterally symmetrical ancestors, Williamson now believed the ancestors of the entire phylum of the echinoderms had been protostomal and pentaradial throughout their original evolution, only later acquiring the bilaterally symmetrical larvae in their development. The acquisition of larvae from a bilaterally symmetrical deuterostome lineage had masked their original evolutionary trajectory and confused the taxonomy of the echinoderms for one and a half centuries. It was a bold, and highly controversial, interpretation.

Williamson now revised his paper and sent it to the marine ecologist David Cushing, a Fellow of the Royal Society, asking the distinguished scientist if he had any idea where he might get it published. Cushing sent copies to two other fellows, after which he returned a complimentary opinion, voiced by all three readers, and offered some constructive editorial suggestions. One of these fellows, a member of the Royal Society's editorial board, expressed the opinion that the paper was 'too philosophical for any of the Society's journals'. It seemed ironic to Williamson that these journals included *Philosophical Transactions of the Royal Society*. The fellow suggested that Williamson consider submitting his paper to *Biological Reviews*. In May, undaunted, Williamson submitted the paper to the *Oxford Surveys in Evolutionary Biology*, a journal with the specific aim of publishing 'new ideas on evolution and to promote controversy'. It was once more rejected by the editor, who commented, 'It is controversial ... not really the sort of paper we had in mind'. That September, and again the following March, Williamson's paper was rejected in turn by the *Journal of Natural History*, *Biological Reviews*, and the *Journal of Zoology*.

Science, of course, is rightly conservative and adheres to principles that have withstood the test of time. But the publication of a new theory does not assume it has been cast in stone: merely that it has sufficient credibility to be considered on its merits by a wide and diverse readership. Despite two years of persistent rejection of his paper, Williamson had continued to break new ground in his experiments, and was increasingly convinced that hybridisation was an important source of evolutionary innovation. Why then was he experiencing this wall of rebuff?

There are a number of possible reasons, including differences of personal opinion and plain incredulity, but over and above all such considerations looms an explanation that seems overwhelmingly likely. As Williamson put it himself: 'When Darwin published his *Origin of Species* in 1859 it was considered heretical by most people. Today the situation is reversed, and, in biological circles at least, it is considered heretical not to agree with Darwin'.[10] But what, then, was the "heresy" Williamson was perpetrating?

Williamson is hardly disputing the basis of Darwinian evolution. There can be no doubt that he is also a serious and reputable biologist. At the time he was fruitlessly submitting his paper, he had published more than seventy original scientific studies covering research into marine biology, and into planktic larvae in particular. He remains a distinguished world authority in his field. It is tempting to assume that no heresy appears minor to a true believer, but that would in turn be unfair to the journals involved and to the integrity of the referees who had turned down the opportunity of publication. Quite simply, Williamson's theory challenged one of the central pillars of biological orthodoxy. This pillar, which is the basis of the evolutionary tree of life, is known as the "biological species concept", or BSC.

Most people feel a sense of awe at the diversity of nature. By diversity we mean that there are a vast number of different life forms, each belonging to its own group, or species. Any individual species is regarded as distinct from other species, to the extent that individuals from one species typically cannot sexually reproduce with another. For example, it is perfectly obvious that a bird cannot cross-reproduce with an earthworm or an elephant with a housefly. Biologists call this reproductive isolation. And reproductive isolation has been regarded, since the time of Darwin, as a key aspect in the origin of new species. But in cases of similar species belonging to a common group, or genus, the degree of distinctness, or reproductive isolation, may not always be as clear as this. Appearances are not everything. At one time a yellow-rumped warbler species that inhabited the western United States was known as Audubon's warbler. This bird has a yellow throat, which distinguishes it from the eastern warbler, known as a myrtle warbler, which has a white throat. Later, it was discovered that the two supposedly distinct species interbred with each other over a wide geographic territory, stretching from Alaska to southern Alberta, so that today the Audubon's warbler and the myrtle warbler are regarded as subgroups within the same species rather than as separate species.

By and large, species do keep to themselves, and thus species tend to follow their own linear trajectories in their evolution. Not only do different species avoid cross-breeding with each other, there are also important physical and genetic differences between them that help to preclude this. While there is an active debate on which takes priority, the reproductive isolation leading to genetic differences, or the genetic differences leading to reproductive isolation, the concept of distinct and separate species, exceptions such as Audubon's warbler aside, is fundamental, not only to the Linnean classification of life but also to the core concepts of neo-Darwinian evolution. Modern Darwinians will often use the term "speciation" as synonymous with the mechanism for the origin of new species, implying that this is the exclusive means of generating new species from old. Any theory that proposes the crossover of genes, including developmental control genes, or, most spectacularly of all, whole genomes between different species, challenges these core beliefs. It is hardly surprising that Williamson was experiencing difficulty in convincing reviewers of the credibility of his theory.

On the other hand, science, by definition, is the pursuit of knowledge and for knowledge to progress, scientists, however sceptical, must be prepared to consider new ideas. In 1987, and quite unexpectedly, Martin Angel, the editor of a journal called *Progress in Oceanography*, contacted Williamson and expressed an interest in his theory. Angel had heard Williamson talk on his larval transfer theory to the Challenger Society for the Promotion of Oceanography and he had seen an early draft of the paper. As a fellow planktologist, Angel sympathised with Williamson. He frequently came across anomalous larval forms himself and he was puzzled by the taxonomic relationships being put forward by colleagues who refused to question conventional explanations. Angel also knew about Williamson's difficulties with the other journals. He declared flatly, 'I'll publish your theory in my journal'.

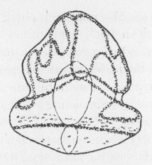

*Tornaria larva of an acorn worm*

5

# Challenging the Tree of Life

WHEN HE PUBLISHED Donald Williamson's paper that
same year, Angel introduced it with what must be one of the
most unusual editorial comments ever included in a modern
scientific journal:

> *It is sometimes stated that papers involving ideas are more
> difficult to get published than those that follow the conven-
> tional wisdom of the time. Iconoclasts are rarely popular with
> their peers. Here is a paper that I am including in Progress
> in Oceanography as something of an editorial indulgence, but
> I am confident that in time the decision will be vindicated.*[1]

Angel went on to list all of the paper's refusals and rejections. From the content of some of the reviewers' comments,
he was convinced that prejudice had blinded them to the logic
of Williamson's arguments. 'Darwin', he wrote, 'would have

probably had less trouble submitting a draft of the *Origin of Species* to the Bishop of Oxford'.

Williamson was at last in a position to tell his colleagues why he disagreed with the prevailing view of animal life history, in which larvae were assumed to fit the same linear evolutionary development as their corresponding adults. It was a risky if courageous undertaking.

Over the century and a half since Darwin introduced his theory of evolution, leading authorities from many countries had used larval features as their guide to fitting entire classes of life into positions in the evolutionary tree. Such relationships were typically depicted as a linear development, in which new life forms exclusively arose through branching from pre-existing species. The implications for larvae, and the evolution of metamorphosis, were evident: the various stages of development, from fertilised egg, to embryo, through one or more larval stages and on to the adult life form, could only have evolved as part of this linear "descent-with-modification" process.

Such a "linear-with-branching" explanation had an understandable appeal in logic. Moreover, as Darwin himself had pointed out, it allowed for the evolution of considerable variety at the larval as well as the adult stage since each stage of the metamorphosis was itself subject to natural selection. In most of the commonly described life histories, the larval form was an adaptation to its free-floating planktic ecology just as the adult form was an adaptation to its more sedentary ocean-floor ecology. Sometimes the adults showed more evolutionary experiment than larvae; in other cases the larvae were the main experimenters. But in one respect all were agreed: metamorphosis through its various stages was so radical a development that it had to carry very powerful evolutionary advantages for it to occur and persist.

The obvious and over-riding advantage for a larval stage was mobility. Surface currents swept those "wanderers" far and wide, vastly extending the geographic and ecological range of the species. All of this Williamson accepted. But it hardly explained why creatures from the same genus, family, class, or phylum, whose larvae occupied the same ecology, should sometimes have very different larvae. No more did it explain why the same larval forms cropped up in the life history of animals from different genera, families, classes, or phyla, with no intermediate forms bridging the evolutionary gap, any more than it explained, to his satisfaction, why some life forms exhibit two or more quite different larval phases, with all the waste, danger, and expended energy this entails, while inhabiting the same planktic ecology. In his paper, Williamson suggested that all such puzzles could readily be explained if large amounts of genetic material had been transferred between different species at some stage in their evolution. Take, for example, the anomaly of the sponge crabs. These true crabs have very similar larvae to the very distantly related hermit crabs. This made perfect sense if the earliest sponge crabs had no larval phase but acquired this later from a hermit crab by genetic transfer.

Why should Williamson devote such intense attention to a seemingly obscure bywater of development? In fact, his purpose was the illumination of the great mystery of how, from this single great and wonderful phylum, Echinodermata, bilateral larvae gave rise to radially symmetrical adults. As Williamson remarks, 'biologists have been asking this ever since the concept of evolution became accepted'. If he was right, the ancestral echinoderms had been radially symmetrical throughout their evolution. This would explain the development of Kirk's brittle star and a number of other supposedly anomalous developments among the echinoderms. After first evolving into the major classes of starfish, sea urchins, and the like, echinoderms had, at some time in the distant past, acquired the genetic template for their bilaterally symmetrical larvae from another species.

This could only be seen as a major evolutionary event.

Darwin shared Haeckel's view that the development of an animal from egg to adult, known as their ontology, holds irrefutable evidence of its past evolutionary pathway, known as their phylogeny. Again and again, Darwin emphasised the profound significance of these early stages of development. 'On this view', he wrote, 'we can understand how it is that, in the eyes of most naturalists, the structure of the embryo is even more important for classification than that of the adult'.[2] Today we regard metamorphosis as a continuation of embryonic development outside the egg, embracing its startling changes within the umbrella of post-embryonic development, or "reprogramming". To give him his due, this was a view that Darwin himself anticipated. Not only did he devote a major section of *Origin of Species* to a detailed discussion of embryonic development, including metamorphosis in insects and marine animals, he explicitly acknowledged that an understanding of metamorphosis was '… one of the most important subjects in the whole round of natural history'.[3]

Well aware that the striking changes seen in metamorphosis might weaken his argument for a slow and linear change under the influence of natural selection, Darwin downplayed their challenge, declaring that the changes seen in the more dramatic cases of metamorphosis, for example from the caterpillar to the butterfly, were unrepresentative. 'The metamorphoses of insects, with which every one is familiar, are generally effected abruptly by a few stages; but the transformations are in reality numerous and gradual, though concealed'.[4] To grasp his point, we might look to the example of a secretive group of creatures, acorn worms, whose ecological importance is easily overlooked.

The acorn worms are not so familiar as the starfish or sea urchins. These ecologically important creatures inhabit U-shaped tunnels in the seabed, constructing tubes of mud or sand particles around themselves and, much as earthworms on dry land, vacuuming organic material from the sand debris of the world's shorelines. While the casual beach walker might not be familiar with the worms themselves, he or she could hardly miss the characteristic coils of sand discarded from the anal end of the tube. Acorn worms also resemble earthworms in shape though not in size – some of them grow to almost two and a half metres in length. Their internal structures also set them far apart from earthworms on the tree of life: behind the proboscis is a cylindrical collar that conceals the mouth, and, in the main body behind the collar, are gill slits opening into the throat cavity. These features of the acorn worms, together with their two nerve cords, place them as a separate class within the phylum of hemichordates, which places them within the superphylum of the chordates, along with us.

The acorn worm larva is known as a "tornaria" because, in shape and the blur of its rotatory movement, which is readily visible under a dissecting microscope, it looks like a child's spinning top. In its subsequent metamorphoses, this larva undergoes considerable changes in shape and internal structure to become the greatly elongated and burrowing worm-like adult. However, all these changes are gradual and linear, much as Darwin envisaged. The front part of the larva elongates into a proboscis, then a collar forms behind this and the remainder of the larval body elongates into the multi-gilled trunk. The respective parts of the juvenile body develop from the corresponding larval structures and the only larval features to be discarded are the bands of locomotory cilia. Indeed, for Williamson at the time of drafting his first paper, it was reasonable to suggest that a linear process of evolutionary change involving descent with adaptation might well explain the sea acorn metamorphosis.

Darwin's linear and gradualist explanation had been accepted and endorsed by generations of evolutionary biologists. But now, in putting forward a theory of evolutionary change that involved transfer of whole genetic programming between dissimilar species, Williamson was contradicting such orthodoxy. The introduction of a novel developmental programme into an existing species would bring about an abrupt and major evolutionary change – what biologists call a saltation. The evolutionary consequences could hardly be described in terms of a branching tree. On the contrary they implied that different branches would recombine, to create, in a single dramatic event, a new life form.

Returning to the larval forms of acorn worms and echinoderms, we encounter an anomaly, however. In 1881, the Russian zoologist Élie Metchnikoff, famous for the discovery of our phagocytic white blood cells, drew attention to the fact that the acorn tornaria also resembles the larvae of certain echinoderms, in particular the starfish. So confusing is the similarity that nineteenth-century biologists mistook the acorn worm larva for that of an unknown family of starfish. During their respective developments from the fertilised egg, the echinoderms and acorn worms share not just the form of the larva, they also share the same process of internal body cavity, or coelom, formation as well as a radial form of early cell division. And so close are these resemblances that contemporary biologists have wondered if both the echinoderms and acorn worms might have evolved from a common ancestor, perhaps one that resembled a tornaria. Any such link depends entirely on the larval similarities.

At the same time, the marked differences between the adult forms are very difficult to reconcile with these larval similarities. This is why Barry Fell suggested that biologists should ignore the larval developments and consider only the adult developments when working out evolutionary relationships. The distin-

guished biologist Pat Willmer, of t
agrees with Fell: 'Very considera
the diversity of echinoderm larva
of echinoderms based on larval f
adult morphology and with fossi
ion, the similarities between the
gent and 'perhaps even meanin

But convergence usually denotes a superficial
Williamson points out that in a number of complex internal
developments, the tornaria really does appear to display the
same origins and patterns as, say, the two-ringed "bipinnarial"
larva of the starfish or the ear-like "auricularia" larva of the sea
cucumber. In his view, this makes convergence unlikely.

Instead, Williamson proposes that we should not ignore the
larval forms – rather we should draw a different, indeed a revolutionary, conclusion. If we accept that the echinoderms and
hemichordates evolved independently and only became related
when a hybridisation event took place between early species
of the two phyla, the adult and the larval forms are illuminating. The common larval forms no longer require some bizarre
switch from bilateral to radial symmetry in the history of the
echinoderms, and given the mechanics of broadcast spawning in the oceans, the opportunity for hybridisation between
different species would have arisen many times. Each putative
hybridisation event would have led to a specific saltationist
genomic experiment.

Williamson conjectures that many such genomic experiments would have taken place over the vastness of evolutionary
time, but the majority probably ended in failure. All it would
take is an occasional successful outcome for the hybrid offspring
to possess the genetic programming for two very different body
plans. Any potential offspring would inherit the developmental blueprints for two distinct life forms, a simpler one for the
planktic life cycle of the larva and a more complex one for the

...ce on the bottom of the ocean. The planktic phase,
...hypothesis, evolved to the present day larvae.

...en the close resemblance between the larva of the sea
...umbers and the tornaria of the sea acorns, it seemed likely to
Williamson that it had resulted from a hybridisation of an early
echinoderm with an ancestor of the present-day acorn worms.
This would offer a logical explanation for how the echinoderms
acquired their first larva. From that abrupt evolutionary event,
over the millions of years since, the genetic blueprint for this
and other related larvae evolving from it, spread throughout
the classes of echinoderms.

But now, a single courageous editor had taken the chance and
published Williamson's theory, making it available for due consideration. If he was right, the conventional theory of linear
descent had so blinkered scientists' vision that it had shifted
the entire phylum of the echinoderms from their proper position on the evolutionary tree. These species were not deuterostomes, like the vertebrates, but members of the opposite great
trunk, the protostomes, which included the superphylum of the
arthropods, such as lobsters, crabs, shrimps, insects, and spiders. Indeed, if his biological colleagues accepted his line of
argument, the elegantly simple tree of life would have to be
chopped down at its first great fork, and a new, more complex
tree would have to be drawn up, with the echinoderms transferred from the deuterostomes to the protostomes, and with
some branches and twigs re-uniting as well as dividing.

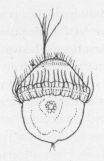

*Trochophore larva of a mollusc*

# 6

# 'This Is Impossible!'

IN THE SPRING OF 1988, Lynn Margulis, Distinguished Professor at the University of Massachusetts, received an unsolicited reprint of a scientific paper in her mail. Arriving out of the blue, and without an accompanying letter, it outlined a remarkable and completely novel theory about the putative role of larval transfer in marine metamorphosis. It was written by a man she had never heard of, a marine biologist named Donald Williamson working in some obscure laboratory in the far-off Irish Sea. She read the paper and was deeply impressed. She wrote back, 'Who are you? I've never heard of you. What a marvellous theory!' Williamson's reply was typically humourous: 'There's no reason why you should have heard of me. I'm not an evolutionary biologist. I've been a planktologist all my life so there's no reason why you should know me … I am sixty-six years old, from a family whose members are short-lived and thus I'm on a straight-line course for posthumous recognition'.

In retrospect, Williamson's approach to the American scientist would prove to be important in obtaining recognition for his theory. In January 2001, when I met him for the first time at his home in Port Erin, I asked Williamson why he had taken this step at that particular moment in his life.

'I knew,' he explained, 'that she had a theory about endosymbiosis as the origins of cells, which, in some ways, was similar to my larval theory. So I thought she might be interested.'

Today Lynn Margulis is a famous figure in the world of biology. A member of the US National Academy of Sciences, she has been honoured with many international awards, including the National Medal of Science. But in 1970, when her book *Origin of Eukaryotic Cells* was first published, she was a youthful thirty-two, and the mother of young children, working as a relatively unknown assistant professor in the Department of Biology at Boston University.[1] Her "serial endosymbiosis theory", or SET, proposes that the first eukaryotes − life forms with nucleated cells − came into being through the symbiotic union of bacteria-like ancestors to create a new composite being, one that ultimately led to the kingdoms of plants and animals as well as fungi and many of the single-celled creatures known as protists. Margulis had endured a protracted battle with scientific orthodoxy before her theory became widely accepted. Now it was Williamson's turn to battle those same conservative forces with a new theory of evolution, based not on symbiotic union but on hybrid sexual crossing. As generous of spirit as she was erudite, Margulis had the wisdom to consider a revolutionary new theory with an open mind, coupled with the strength of will to combat the inertia and antipathy of her more conservative colleagues. In reading Williamson's paper, she could only be transfixed.

In 1988, Williamson submitted a new paper on larval transfer to the two leading scientific journals, *Nature* and *Science*. Both journals rejected it. In this paper, Williamson had described how he had fertilised sea squirt eggs with sea urchin sperm, resulting in sea urchin pluteus larvae. The referees declared that this was impossible. Margulis suggested he write a shortened paper for the reputable American journal, *Proceedings of the National Academy of Sciences,* or PNAS. Williamson now modified his paper to suit that journal's requirements. The editor sent it out to four reviewers, whose responses proved 'inconclusive'. More opinions were sought. One reviewer point-blank refused to read the paper, branding Williamson's ideas 'high school level rubbish'. Another thought the fossil record was consistent with Williamson's ideas, but he was unable to evaluate the article's developmental arguments because the territory was highly specialised even for an experienced biologist. A third commented that Margulis should end her support for Williamson forthwith before her scientific reputation was jeopardised. A consistent criticism was the need for better photographic documentation of the hybrid organisms. Margulis concurred. Others thought, not unreasonably, that molecular studies were needed to confirm or refute Williamson's claims. At the same time, nobody could deny that his theory was abundantly clear in its proposals and should be readily testable by experimental confrontation and by standard molecular biological investigation.

While the paper was still undergoing peer review, Margulis introduced Williamson's ideas to Alfred I. "Fred" Tauber, MD, Professor of Medicine at Boston University. Tauber also served as the director of the Center for the Philosophy and History of Science at Boston University and was arranging a symposium on the subject of "The Self". When Tauber asked Margulis for names of appropriate speakers, she had no hesitation in recommending Williamson: his theory challenged the very notion of self-hood. In Tauber's words, 'Lynn recommended Williamson and the controversy surrounding his fascinating work'. But Tauber understandably felt obliged to question what the controversy really amounted to. Was Williamson being 'very clever but deceptive'? Or was he proposing one of the most interesting ideas in a century involving marine zoology? Could he convince a sophisticated audience that his theory was based on adequate observation and sound scholarship?

Ahead of the symposium, and during a coincidental trip he had already planned to the United Kingdom, Tauber detoured to the Isle of Man, meeting up with Williamson in the Marine Laboratory. It is evident from his subsequent comments that Tauber's fears were put to rest. In Williamson he believed that he had come across one of the last great nineteenth-century-style naturalists. Writing back to Margulis, he confirmed the marine biologist was an erudite and careful scientist whose originality merited airing and appropriate criticism. Williamson was well known as far afield as Japan and Korea for his previous work on shrimps and other crustaceans. By then retired from his position at the Marine Lab, he continued to work in an emeritus capacity, with access to laboratory and office accommodations, but without genetic and molecular biological support. 'Although he has little in the way of modern technology, he has performed decades of careful work with equipment at least as good as that of Ernst Haeckel ... on a theory based solely on observation, intuition, a broad grasp of phylogenetic relationships, and the tantalizing first experiments that appear to support his argument'.[2]

Williamson felt encouraged to continue with his work. In 1989, he repeated the cross-phyletic experiment, breeding eggs of the sea squirt *Ascidia mentula* with the sperm of the sea urchin *Echinus esculentus*. The new experiment was spectacularly successful and he obtained more than 230 fertilised eggs, a great many of which developed into *pluteus* larvae typical of the paternal sea urchin. What was this if it was not larval transfer! He ran parallel control experiments in which he fertilised sea urchin eggs with sea urchin sperm, and these produced indistinguishable pluteus larvae, which developed at the same rate. With a growing excitement, Williamson followed the progress of his test larvae. Through the low-power binocular microscope, he watched them swim in the seawater of his culture dishes with a whirring border typical of surface cilia. Larvae of the maternal ascidians are not ciliated. He observed how they grew progressively until they were about a millimetre long and just visible to the naked eye. This was the stage when they should be developing into juvenile adults – but through all of his patient observation they failed to do so. 'I kept them for over a month, with some of them dying off all the time, and eventually they all died without any of them metamorphosing'. This endpoint to the experiment was bitterly disappointing, though to a detached observer Williamson's results in such few experiments over such a short space of time would appear truly astonishing. 'At this time, I was preparing a more comprehensive paper on this experiment. I knew, of course, that I would have difficulty getting anybody to believe me'.

On March 30, 1990, Williamson travelled to Boston at Margulis's invitation. He had the pleasure of meeting the distinguished biologist for the first time and of presenting his theory at the Boston Colloquium for Philosophy of Science. The title of his talk was "Sequential Chimeras". He took the stage before a gathering of some thirty people in the cavernous Boston University Law School auditorium, built to house an audience of five hundred.

Quoting Homer, he explained his use of the term "chimera", which derived from the Greek fable of a fire-sprouting monster, with a lion's head, a she-goat's body, and a serpent's tail. The word had been adopted in more prosaic fashion to describe the experimental blending of tissues from two different species, or of two genetically distinct strains of the same species, by techniques such as grafting or genetic engineering. 'According to most dictionaries', he noted, "chimera" has another definition, one often regarded as the most common usage, where it refers to a mere wild fancy, an unfounded conception, an absurd creation of the imagination'. He added, with a chuckle: 'I should now like you to consider the possibility, indeed the strong probability, that organisms with identifiable features of different origin existed long before there were genetic engineers or even ancient Greeks'. In this way he introduced them to the idea that a great many animals (and plants too, though he wasn't to address this in the lecture) are really chimeras, produced by nature over the vast periods of evolutionary time. Unlike the grafted rose or the mythical monster, these chimeras did not readily reveal their miscegenetic origins; rather, their blended ancestry manifested itself, like successive acts of magic, during consecutive stages of their life histories. He felt it not unreasonable to refer to them as "sequential chimeras", since they began with the features of one group of animals before metamorphosing to an entirely different group, frequently a different phylum. Indeed, if his reasoning were to be accepted, Haeckel's elegant evolutionary tree was about to become extinct.

His audience responded with predictable incredulity. Like Williamson himself, they had been steeped in orthodox views since their school-day biology lessons. They were well aware that Darwin saw larval forms as important clues to understanding the evolutionary past. For Darwin, the conclusion was inescapable: 'As the embryo often shows us more or less plainly the structure of the less modified and ancient progenitor of the group, we can see why ancient and extinct forms so often resemble in their adult state the embryos of existing species of the same class'. In *Origin of Species*, Darwin had illustrated this with 'what we know of the embryos of mammals, birds, fish and reptiles – that these animals are the modified descendants of some ancient progenitor, which was furnished in its adult state with branchiae [gills], a swim-bladder, four fin-like limbs, and a long tail, all fitted for an aquatic life'.[3] On this point Williamson agreed with Darwin. The development of echinoderms and other phyla really did provide clues to their evolutionary development and taxonomic relationships. 'But if we are to accept Darwin's viewpoint, echinoderm larvae have left-right symmetry because, at an early stage in the evolution of the entire phylum, the adults also had left-right symmetry. This leads us to believe that the larvae have retained this while some of the adults have evolved radial symmetry. Meanwhile we have to presume that others, which might have retained the left-right symmetry, have become extinct'.

This was what his audience actually believed. But then he went on: 'I subscribed to this myself for many years and taught it to my students, but now I have changed my mind'.

He invited them to consider the fossil record of the echinoderms, which was plentiful despite its great antiquity. Many bizarre and beautiful early forms had been found, some resembling plants, complete with stems and roots, or spiral shells, or even hand-held mirrors – like the most wonderful experiments in three-dimensional sculptural form. But while a proportion of these had three arms and some five, none were bilaterally symmetrical. To this Williamson added that no larval forms have been found in echinoderm fossil strata Cambrian deposits, dating to more than five hundred million years ago. Though biologists might assume that this was the result of their soft bodies leaving no fossil impressions in rock strata, other soft bodies had left striking impressions. Yet this lack of a fossil record fitted perfectly with his proposal that the echinoderms had not acquired larvae until the Carboniferous period, some two hundred million years later, a time when echinoderms indisputably were radially symmetrical. To put it bluntly, the conventional hypothesis for the evolutionary origin of echinoderm metamorphosis was, when one examined it closely, fraught with difficulties.

'I'm well aware', he confessed, 'that from the point of view of the Darwinians, this makes me a heretic. But mine is a relatively small heresy. I agree with Darwin that organisms have evolved from other organisms and that natural selection has played an essential part in determining survival or extinction. I also agree with him that evolution is not restricted to adults: embryos and larvae too have evolved and are still evolving. I have, however, come to disagree with his view that embryos and larvae must, in all cases, have the same genealogy as the adults they give rise to'.[4]

Williamson would subsequently deliver this same talk to members of the Marine Biological Laboratory at Wood's Hole, Massachusetts, one of the leading research facilities in the field. Speaking calmly and elegantly, and peppering his talk with humour, he hardly cut the figure of a heretic. But in Tauber's memory, 'To the experts who were present, Don's hypothesis was far-fetched, if not simply outrageous'.[5]

Williamson reminded his second audience that he was not the first to adopt such heretical views. Fell, the eminent New Zealand expert on echinoderms, had arrived at the same conclusion as long ago as 1948. Observing that larval morphology was full of incongruities, Fell recommended the very opposite of Darwin – that biologists should focus on adult morphology alone, and disregard the larval evidence, when deducing how the phylum had evolved. While Fell had made no convincing attempt to explain the incongruities between adults and larvae, Williamson offered a logical and consistent synthesis. 'I postulate that the first echinoderm larva originated in another phylum, that its acquisition by an echinoderm did not occur until after all the existing classes were well established, and that the larva gradually spread from one individual to another, species to species and group to group within the phylum, evolving as it spread'.

Williamson illustrated what he meant with the example of *Astropecten auranciacus,* a gold-coloured starfish that inhabits coastal waters, and one that was very familiar to the marine biologists assembled.

'The adult is a five-armed star that crawls over the ocean floor. It is radially symmetrical. The larva on the other hand is bilaterally symmetrical. Why, then, should *Astropecten*, like thousands of other echinoderms, spend the early part of its life as a bilaterally symmetrical larva and then switch, dramatically and totally, to radial symmetry for the remainder of its adult life?' What, in other words, was the advantage of bilateral symmetry to a free-floating organism? Would it not have made more sense the other way round, with the radial animal floating freely and the bilaterally symmetrical animal crawling over the ocean floor? He showed his audience a picture of the *Astropecten* larva, which looked, in silhouette, like a sport winner's two-handled trophy cup. The larva is classified as a bipinnaria, possibly from the fact the handles of the cup are like two locomotory wings, rather as we saw with the larva of *Luidia sarsi.* 'This', he explained, 'is the typical larval form of the majority of starfish. According to Darwin, any other marine creature with this type of larva must share a common earlier ancestry with the starfish'. He projected onto the screen the very similar tornarial larva of the acorn worm. In the acorn worm's phylum, as Williamson freely admitted, the development from larva to adult exhibited none of the catastrophic violence seen in the starfish. 'There is no quasi-parasitic behaviour. Instead, the adult really does develop from the juvenile, much as a human foetus does from the embryo, by a process of extension and differential growth'. Here were two very similar, and potentially related, larvae, each participating in the metamorphic life cycle of very different phyla, experiencing very different patterns of development. In Williamson's opinion, this made no sense when interpreted through the orthodox evolutionary perspective.

He shifted ground to pose a hypothetical question. What if there were no larval similarities between the echinoderms and hemichordates? Would two such dissimilar phyla ever have been linked to the same branch of the evolutionary tree? Of course he knew that they would never have been linked. 'It is time', he concluded, 'that we chopped down this phylogenetic tree, since the clues from embryos and larvae have been just as badly misinterpreted in the other main branch'.

Biologists needed a new theory, he suggested, one that reconciled the similar larvae as well as the extremely dissimilar adults of echinoderms and sea acorns: his own theory of larval transfer.

He began with the echinoderms. Originally this phylum had no larvae. They were radially symmetrical throughout life. At some subsequent stage the echinoderms incorporated the genetic prescription of a bilaterally symmetrical animal, such as the acorn worm – an ancestor that subsequently developed to a larval stage similar to an acorn worm's tornaria. This led to echinoderm metamorphosis, which accommodated two developmental blueprints. Where the acorn worm plan was bilaterally symmetrical throughout the life cycle, the echinoderm metamorphosis inserted the prescription for a bilaterally symmetrical life history into a pre-existing blueprint for radial symmetry.

The gastrula stage, where the embryo had evolved the beginnings of a mouth and gut, was the key to this development, and to the configuration of the cells that would become the newly forming echinoderm. In Williamson's opinion, the early echinoderms, without larvae, would have developed their pentaradial shape from the blastula, as seen in Kirk's brittle star. After an echinoderm and a hemichordate hybridised, the hemichordate portion of the genome took control of the first half of development, producing three pairs of coelomic sacs, all composed of stem cells. When one of these sacs reached the right size, the echinoderm portion of the genome recognised it as similar to a blastula, and it proceeded to make a pentaradial echinoderm from it. In this way, the first echinoderm to acquire a larva added a bilaterally symmetrical planktic phase to its life history. This gave many echinoderms an adaptive advantage over their non-metamorphosing relatives: wider dispersion as bilaterally symmetrical plankton allowed for easier escape in times of starvation or environmental stress. Natural selection would have selected for the metamorphosing species.

This concept of whole genomic incorporation, with its potential for dramatic metamorphosis between life-cycle stages, Williamson dubbed his "sequential chimera theory". He was unapologetic in affirming that his theory of the acquisition of larval genomes was an example of non-Darwinian evolution, involving a reticulate or blending pattern of evolution as opposed to the classical linear, branching pattern, and that it assumed a sudden evolutionary event as opposed to the classical gradualist pattern. Indeed, he speculated that this evolutionary mechanism went much further. Animals in many other phyla have incongruous larvae, which suggest they too are later additions to the life histories in which they now occur. Nowhere was this more apparent, or interesting, than in another familiar larva, the trochophore.

To a non-professional eye the trochophore somewhat resembles an acorn with a furry hatband and a Mohican-style apical tuft of hair. In fact the term "trochophore" is taken from the Greek for wheel, since the furry hatband is, in reality, a tyre-like equatorial ridge, bearing a ring of locomotory cilia, which causes the larva to spin like a wheel as it moves through the planktic waters. Trocophores are found in four different phyla, including annelid worms, spoon worms, peanut worms, and molluscs. Molluscs are the third largest phylum among all animals, with some fifty thousand species, including snails, slugs, clams, cuttlefish, and octopuses. They are among the most important and varied of the marine invertebrate animals, dating back to the earliest Cambrian strata, perhaps even earlier still, to the first soft-bodied animals of the so-called Ediacaran period, between 570 and 600 million years ago. Williamson drew the attention of his audience to the fact that two whole classes of molluscs have no larvae, notably the cephalopods, including the octopuses, as we saw earlier. Meanwhile, trochophores are found in some but not all of the remaining classes of molluscs, and variants of the trocophore are also found in the phylum of the nemertine worms. In fact, trochophore variants occur in at least seven different marine animal phyla, where some metamorphose to juveniles while others metamorphose to another, radically different, type of larva. Did this imply that all of these phyla shared a common ancestor? And if so, how does one explain that yet other molluscs develop through quite a different larva, known as a pericalymma?

Williamson went on to describe the catastrophic changes that accompanied metamorphosis in many marine animals, where independent development of the juvenile adults was reminiscent of what was seen in the echinoderms. To him, many of these observations made no sense from the perspective of linear descent with modification. But if his theory was right, it offered a rational explanation for such anomalies, with implications every bit as startling for genomic evolution as witnessed in the gross physical violence seen in the body and tissues of the animals. He did not deny that his theory invoked difficulties that needed to be overcome – difficulties of both a theoretical and practical nature. 'I admit', he concluded, 'that the full implications cannot be assessed until the extent and limitations of larval transfer are defined'. In particular the genetics of such hybrids needed to be evaluated. Complex genomic rearrangements must come into play, especially so when the parental generations were widely separated in the evolutionary tree. What form these rearrangements would take depended on genomic mechanisms he could neither imagine nor predict.

Williamson's lecture fell like a bombshell upon his listeners. In Fred Tauber's recollection: 'I rose up and assessed the audience – which included some apoplectic sea urchinologists. Either we were all privy to an historical moment, or else Williamson would be recorded as an ordinarily competent and diligent zoologist, who lived with animals by the seaside for thirty years, and whose mistake should be forgiven in the light of his standard contributions'. In Williamson's more phlegmatic recollection: 'There were a few questions at the Boston symposium but a lot of questions at Woods Hole'.

He came away with an overwhelming impression of disbelief. Several responded with outright rejection of his ideas. 'They told me flatly that they just didn't believe it. It couldn't happen. I was nuts. I must have mixed up my cultures'.

The very referee who had turned down his submission to the *Proceedings of the National Academy of Sciences* as not worthy of a competent sixth-former was in the audience. He now took to his feet and confronted Williamson:

'This is impossible'.

Williamson would subsequently recall: 'His name was Dick Whittaker. He was a marine biologist doing research on ascidians and he knew a lot more about ascidians than I did'.

'Very well', Williamson countered. 'This is a marine station. Presumably you have sea urchins and ascidians here. Let's try the same experiment here and now. Let's start tomorrow morning'.

J. Richard "Dick" Whittaker was surprisingly game. He duly produced some urchins and ascidians. But when they attempted a new cross-phyletic hybridisation, they obtained no living offspring. Williamson perceived, however, that Whittaker was finally convinced that he was not talking nonsense. One or two eggs divided once, but then they stopped. This suggested to Whittaker that it could just possibly happen.

Williamson's American lecture was published as a chapter, predictably entitled "Sequential Chimeras", in a book edited by Tauber, *Organisms and the Origins of Self*. In the introduction, Tauber commented that the incredulous reception Williamson received in Wood's Hole was only to be expected. But the very essence of science was the development of and testing of ideas and theories, and Williamson's theory was at the very least stimulating debate and experiment. Tauber summarised a number of questions that needed to be answered. More photography and anatomical dissection of the hybrid offspring was needed. Ascidians were hermaphrodites and might thus self-fertilise. Future experiments would need to rule this out with particular detail. Williamson might well counter that the offspring were not those you might expect of an ascidian mother, whether through hermaphroditic or sexual fertilisation, but of the echinoderm father. But there was no doubt about the relevance and importance of Tauber's final suggestion. Detailed genetic confirmation was essential if the hybridisation results were to be believed.

Williamson returned to the Port Erin Marine Laboratory in early May 1990, determined to resolve these questions.

*Metamorphosis of hybrid pluteus to spheroid*

# 7

# *Catastrophe*

IN FEBRUARY 1990, and prior to his trip to America, Williamson had initiated a new cross-phyletic hybridisation experiment, fertilising another batch of eggs from the sea squirt *Ascidia mentula* with sperm from the sea urchin *Echinus esculentus*. In keeping with his increasing expertise in hybridisation methods, this experiment yielded an enormous number of successful hatchlings, three thousand of which would successfully metamorphose. As previously, the hatchlings developed not to the tadpole larva of the maternal ascidian but to the paternal pattern of pluteus larvae. He fed the huge harvest on a diet of diatoms and kept them alive in several glass bowls of filtered seawater, changing the seawater twice a week. By March 12, the larvae were a little more than a millimetre long – relatively large for a planktic larva. The majority then underwent a curious change, resorbing their ciliated arms and the internal skeleton that supported the arms. Williamson monitored this change, as they rounded off into spherically shaped offspring that buzzed about the water table through the action of surface cilia. He called these strange new organisms "spheroids". Earlier hybrid researchers had also reported rounded shapes, but these had not developed from plutei and Williamson thought these earlier observations were probably the results of polyspermy, a developmental anomaly resulting from the fertilisation of the ovum by more than one sperm. He also noticed that his spheroids developed a little bump on the surface. This appeared to be a specialised organ, a suction cup that enabled them to attach to the walls of the glass bowls, where they would remain affixed for hours at a time. He magnified the spheroids and took photographs to record their appearance.

When I first saw microscopic specimens and pictures of the spheroids in January 2002, I asked Williamson if these organisms with their ciliated surface and suction cups had ever been reported before. He shook his head. 'No,' he declared. 'These really were new. They were neither tadpoles nor plutei – although they had metamorphosed from plutei. Such things have never been described in echinoderm development at all.'

I was aware that sea squirts, unlike urchins, do attach themselves to rocks and other solid surfaces, though not, to my knowledge, through the action of suction cups. But the tube feet of the sea urchins did employ suction cups.

'Could these spheroids,' I pressed, 'represent the original larval stage of the sea squirts before they acquired tadpoles?'

Again, he shook his head. 'I couldn't determine this one way or the other.'

The majority of the hybrid larvae metamorphosed to spheroids, but in a smaller number – perhaps seventy or so – the pluteus morphology persisted and these developed "rudiments" within their coelomic pouches. In normal sea urchin pluteus larvae, these rudiments would signify the start of metamorphosis to the juvenile adult urchin. In these seventy or so individuals, juvenile adults developed into the familiar pentaradial form. Williamson observed these changes, noting that they followed the normal pattern of echinoderm metamorphosis. He was thrilled when the tiny sea urchin juveniles crawled out from the discarded wrecks of the dead plutei that had been sacrificed for these beautiful yet curiously uncaring offspring.

When Williamson had travelled to Boston, he had left this on-going experiment in the capable hands of his colleague, Alan Bowers. The large numbers of spheroids had been kept in their bowls while the hybrid urchins had been transferred to a small Perspex aquarium tank, kept at fifteen degrees Celsius and continuously irrigated with a trickle of filtered seawater. After his return home, Williamson took up the experiment from where he had left it, moving the surviving spheroids into the same tank as the urchins, meanwhile keeping a few in bowls to see if they would change into anything else. The urchins continued to grow but the spheroids retained their size and form, suggesting that this change was an end point in their metamorphosis. Williamson now counted some twenty free-living sea urchins. He continued to observe both sets of offspring, while collecting more specimens to repeat the experiment. As usual, he was limited by the urchin-breeding season. Meanwhile, he began to write a new paper, intending to report the data he had amassed, while refuting the suggestion of some among his American audiences that he might have mixed up his cultures. He had, of course, taken ample precautions to avoid mixing the cultures and was confident this could not have happened. He also planned some new experiments.

While collecting a bucketful of fresh urchins on a rocky shore, he jumped down from a boulder and slipped on the rocks, striking his head. It seemed no more than a minor accident, resulting in a small gash. He drove back to the lab, where he put the urchins into a seawater tank. As he was doing so, his right arm felt as if it was misbehaving. 'I had some movement but I was not in full control of the movement. My whole arm felt awkward and clumsy. I realised something was wrong. So I headed for home. But along the way I lost the use of this arm entirely and meanwhile I was also losing control of my right leg. The car had an automatic gearbox, so I was able to use my left foot on the accelerator, eventually turning into the drive with one hand. I couldn't complete the lock and ended up stuck half in and half out. I knew by now that I was having some sort of stroke'.

His wife, Enid, rushed to his assistance and phoned the family doctor. By the time he arrived by ambulance into the casualty department of Noble's Hospital, Don could hear and understand what people were saying to him but he couldn't voice a reply. When he did try to speak, his words emerged as gibberish.

The doctors confirmed he had suffered a cerebral haemorrhage. At a critical moment in his hybridisation experiments, the iconoclast had lost his voice.

It was a desperate tragedy for Williamson, lying shocked and dispirited in his bed in the island capital of Douglas, wondering if perhaps he would ever be able to continue his extraordinary line of research. We shall return to the story of this dedicated scientist and his quest. But for the moment let us turn from the exotic world of marine invertebrates to the equally exciting and colourful world of insects, where many distinguished thinkers had long shared the marine biologist's fascination with the strange mystery of metamorphosis.

# Part II
# The Butterfly's Tale

*You have a story to tell. Tell it to me; and for a year, for two years or longer, until I know more or less all about it. I shall leave you undisturbed, even at the cost of lamentable suffering to the pines.*

*But let me tell you: we are called old-fashioned, you and I, with our conception of a world ruled by an Intelligence, we are quite out of the swim. Order, balance, harmony: this is all silly nonsense. The universe is a fortuitous arrangement in the chaos of the possible ... Chance has decided all things ... It hardly seems so. The riddle is as dark as ever.*

Jean Henri Fabre,
'The Life of the Caterpillar'

*Great peacock moth*

# 8

# The Evening of the Great Peacock

IT IS LATE NINETEENTH-CENTURY France, nine o'clock, a warm evening on the sixth of May. At his home, the *Harmas de Sérignan*, the naturalist Jean-Henri Fabre is preparing for bed. Suddenly there is a commotion in the adjacent bedroom, where his son is supposed to be sleeping. Little Paul, half undressed, can be heard rushing around his bedroom, jumping and stamping, knocking the chairs over like a mad thing.

'Come quickly, papa!' he shrieks. 'Come and see these moths, big as birds. The room is full of them'.

Fabre rushes into his son's bedroom to witness an unprecedented event even in his wide experience of the strange and wonderful world of insects. Little Paul's bedroom has been invaded by swarms of gigantic moths. Four are trapped in the birdcage and many others are fluttering their wings against the ceiling.

'Put on your things and come with me!' Fabre exclaims.

In a bustle of activity, Fabre helps his son dress and they hurry downstairs to the study, which occupies the right wing of the house. In passing through the kitchen, they have to skip around the frightened servant, who is flapping her apron at swarms of the same invaders. Grabbing a candle on its sconce, they enter the study, where the scene resembles a wizard's cave. Swarms of the giant moths fill the air, spiralling and eddying like a swarm of bats. With a soft flick-flack they dart and knock around the bell jar Fabre had that very morning placed on his laboratory table. Inside the bell jar is a female moth of the same species as the invaders, a magnificent great peacock. Fabre had watched her emerge from the cocoon he had collected from the bark of an old almond tree. He had cloistered her, still damp with the humours of hatching, under the wire-gauze bell jar, not because he was planning any specific experiment, but merely to indulge his curiosity of nature. But now it looked as if this imprisonment had triggered the male invasion.

The great peacock moth, *Saturnia pyri*, has a broad distribution through the warmer regions of Mediterranean Europe, extending south to North Africa and east to the Middle East. It is spectacularly attractive, with a wingspan that can exceed fifteen centimetres – almost as wide as the open adult hand. 'Who does not know this magnificent moth', writes Fabre of the male of the species, 'clad in maroon velvet with a necktie of white fur, its wings, with their sprinkling of grey and brown, crossed by a faint zigzag and edged with smoky white, in the centre a round patch, a great eye with a black pupil and variegated iris containing successive black, white, chestnut, and purple arcs'. Now, suddenly and unexpectedly, these exotic giants had taken possession of his home. And over the days that followed, the visitors continued to flock to the bell jar, in what appeared to be extraordinary numbers. Fabre would later describe his astonishment in his exquisitely perceptive book, *The Life of the Caterpillar*.

*The aggregate of the visitors during those eight evenings amounts to a hundred and fifty, an astounding number when I consider how hard I had to seek during the following two years to collect the materials necessary for continuing these observations ... For two winters I visited every one of those decayed trees at the lower part of the trunk, under the tangle of hard grasses in which they are clad, and time after time I returned empty-handed. Therefore my hundred and fifty moths came from afar, from very far, within a radius of perhaps a mile and a half or more. How did they know of what was happening in my study?[1]*

With this question began one of Fabre's now-famous experiments, investigating how the male moth tracks down his mate from great distances, negotiating stormy weather and every obstruction. Vision could hardly be the answer. The sexual seduction, and that is exactly what he was observing, took place over too great a distance and in the darkest night. 'No – the path drawn by the male in search of the female is far too erratic for any visual signal, leaving only the possibilities of scent or sound ...'.

Today, more than a century after Fabre made his original observations, his descriptions remain the most enchanting ever made in the world of entomology – and, some believe, in the entire realm of nature. Those wishing to learn more can explore his book, in which one discovers the sole purpose of this sumptuous grace and beauty – the three days of life and obsessive purpose of the adult male moth – is a ceaseless search for the female and the faintest hope of mating.

In *The Descent of Man*, first published in 1871, Charles Darwin, a contemporary of Fabre, pioneered the concept of sexual selection as an evolutionary force.[2] Mammalian females brood their young within their bodies. Even after birth, whether the offspring first enters the world as a still gestating egg or infant, the female is often the parent with the heaviest responsibilities. Females have a vested interest in the genetic health and vigour of their offspring. Evolution is all about passing on the genes of inheritance and nowhere is this more focused than in the act of mating, with its potential for offspring carrying those genes. It is not surprising then that females are rather choosy about their mates. All those wonderful rituals of courtship arise from this: the jousting of stags, the tail of the peacock, the red breast of the perky robin in our garden, and the mimetic song of the lyre bird. Each male endowment is the result of female sexual selection pressure over vast periods of evolutionary time. Yet nature, in its biological strategies, can be altogether ruthless. Some female praying mantises, including a British population of *Mantis religiosa*, will cannibalise their partners, head first, during copulation.[3] This same ruthlessness has condemned the adult male of the great peacock moth to his uncontrollable desire to find a mate, his life encapsulated to a few days of frantic searching. His mouth parts are so rudimentary, he cannot eat or drink. The bulk of his energy, inherited from the dedicated feeding of the caterpillar stage, has been diverted to the production of his vainglorious body, bedecked with the four giant wings, each carrying the distinctive peacock eye.

In his investigation of the male's courtship flight, Fabre paid particular attention to the quadruple antennae, massively enlarged when compared with those of the female. She, on the other hand, 'appears equipped for seduction with a song as irresistibly enchanting as that of the sirens in the voyages of Ulysses'. But was it really a siren song? In fact a song of any sort was soon ruled out of the question. 'The great fat [female] moth, capable of sending a summons to such a distance, is mute even to the most acute hearing', as Fabre now realised. His first big clue came from the fact that he could halt the attraction of the males towards the females by cutting off the male moth's antennae. This suggested that scent might be the answer. 'Are there, in point of fact, effluvia similar to what we call odour, effluvia of extreme subtlety, absolutely imperceptible to ourselves and yet capable of impressing a sense of smell better-endowed than ours?'

Fabre was more than just a poet of nature: he was a master of scientific experiment. And in such 'effluvia of extreme subtlety' he had indeed discovered the solution to the mystery. Today we know that the female great peacock moth attracts the males from a mile or more distance by exuding the equivalent of an extremely subtle scent – her siren call is a chemical of great potency known as a pheromone. This seductant, released by scent-forming patches on her abdomen, has evolved with all of the selfish ruthless of natural selection with the exclusive purpose of eliciting courtship. The male is exquisitely sensitive to her pheromone, which he detects, exactly as Fabre surmised, by using his richly developed antennae.

The pheromones of insects have practical importance. In silkworm moths, for example, where the active substance has been extracted from the tail ends of the abdomens of some half a million females to be isolated in a pure enough state for chemical analysis, it was found to be a simple alcohol, with a sixteen carbon chain. At the slightest contact with this substance, the male violently vibrates his wings. Further experiment has revealed the exquisite sensitivity of the male silkworm moth to this siren scent, when the most minuscule secretion that modern science can detect – defined as a unit of attraction – is in the order of $10^{-10}$ of a microgram. This is one ten thousandth millionth millionth part of a gram. The pheromone is picked up by the antennae of the male and it guides him with uncanny accuracy through dark and storm to find the seductive source. When males of the Chinese saturniid moth, *Actias selene*, were released at a distance of eight miles from caged females, more than one in four of them found their way to the cage. It made no difference, as Fabre himself established, if scientists placed all kinds of additional strong odours as obstacles in their path. The male was not deflected in the least. Today we know that it is not only the female that produces these subtle signals of seduction. In many male moths and butterflies specific scents are liberated by special glands that have an equivalent aphrodisiac function, in turn exciting the female to accept the attentions of the courting male.

Fabre conducted his studies not only on the bench and in the test tubes of his home laboratory but in his *harmas*, an enclosed piece of land behind his house. In this natural sanctuary, he planted suitable shrubs and trees, otherwise allowing nature to run wild. During the summer following the evening of the great peacock he also paid street urchins to collect caterpillars of the moth for him, at a sou apiece. From these he reared his new menagerie on almond-tree branches in his *harmas*, watching them mature and then pupate in the cracks and crevices of the bark. That winter, assisted by his friends, he searched through the brambles at the base of the almond trees – the leaves of which were the chosen food of the caterpillars – to become the possessor of an assortment of cocoons, the bulkier and heavier of which denoted metamorphosing females, which he then would observe for a new generation.

Fabre also dissected some of the stout brown cocoons, so curious with their exit-shaft shaped like an eel-trap, using his pocket knife to cut open the leathery case to see what was happening inside during that long hermitic exclusion from the world. How, he wondered, from this intermediate stage of dormancy, did that extraordinary metamorphosis take place, leading to the emergence of the magnificent adult creature, perfectly formed from the moment of its birth, with its multi-faceted eyes, its plumes of antennae, and its great wings unfurling and stretching in the warmth of the sun, each individual birth a celebration of natural perfection? Did the wingless yellow caterpillar, crowned with its palisade of black hairs and its beads of turquoise blue, invoke some natural magic or spell to grow those enormous wings or those huge fleshy antennae or the compound eyes? What he discovered was a mystery far more beguiling than might be evoked by spell or magic.

The development of the caterpillar really does take us into a realm of the wonderful. Indeed, the apparent quiescence of the pupa could not be more misleading. Inside its leathery wall, the living being of the caterpillar does not change by the addition of wings or eyes to its former body. Instead the body of the caterpillar melts to an organic soup of cells to begin its biological construction all over again.

For the deeply religious Fabre, it must have appeared for all the world as if he were witnessing a miracle of nature in which the newborn insect, with its multi-faceted eyes, articulated legs, new-found sexual maturity, and extraordinary wings, emerges from the sacrifice of its former self in what amounts to the real-life enactment of the mythical phoenix arising from its own funeral ashes. How could such a riddle ever have come to be?

The answer would not come from the gentle Fabre, whose writings were so imbued with an enchantment of description he has been labelled 'the Homer of Insects', but from another genius with an equally profound interest in nature, Charles Darwin. Ironically, Fabre would refuse to contribute to the evolutionary aspects of its ultimate solution, despite being invited to do so by Darwin himself – and the reason for his refusal lay in the very poetic nature of his genius.

Darwin's landmark book on evolution, *On the Origin of Species by Means of Natural Selection*, was published in 1859 – a generation prior to Jean-Henri Fabre's original French publication of "The Great Peacock" in *Souvenirs Entomologiques*. We know Fabre read *Origin* just as we know Darwin read Fabre's *Souvenirs*. Darwin made no secret of his admiration for Fabre, not only for the beauty of his descriptions but for the precision of his experimental method; he referred to him as 'that inimitable observer'. But Fabre did not reciprocate with regard to Darwin's evolutionary theory, a rejection that must have dismayed the ageing naturalist.

In a letter dated January 31, 1880, Darwin wrote to Fabre to suggest, 'I am sorry that you are so strongly opposed to the descent theory; I have found the searching for the history of each structure or instinct an excellent aid to observation; and wonderful observer as you are, it would suggest new points to you. If I were to write on the evolution of instincts, I could make good use of some of the facts you give'.[4]

But even this respectful appeal failed to persuade Fabre, who was irreconcilably opposed to the very concept of evolution.

In chapter eight of his book *More Hunting Wasps*, first published in English in 1919, Fabre described a 'nasty and seemingly futile' experiment in which he reared caterpillar-eating wasps on a 'skewerful of spiders'. To this he added the explanation: 'I should not have undertaken these investigations, still less should I haven spoken about them, not without some satisfaction, if I had not discerned in the results ... a certain philosophic import, involving, so it seemed to me, the evolutionary theory'.

Whereas the diet of the caterpillars was omnivorous, Fabre noted, the parent wasp was limited to a single kind of prey, whether a particular kind of cricket or locust. Assuming that the diet of the ancestor of the hunting wasps was omnivorous, like the caterpillar, he demanded to know why, through evolution – 'that pitiless fight for existence that eliminates the weak and incapable and allows none but the strong and industrious to survive' – the descendant wasps had given up their eclectic advantage to depend on such a limited diet, which must often threaten starvation. In Fabre's own words, 'It is assuredly a majestic enterprise, commensurate with man's immense ambitions, to seek to pour the universe into the mould of a formula … But … in short, I prefer to believe that the theory of evolution is powerless to explain [the wasp's] diet'.

From the specific, Fabre extrapolated the general. Darwin's evolutionary theory was a mathematical contrivance through which 'one was merely observing the same idea from different points of view'. He assumed, incorrectly, that the mathematical reductionism of evolutionary theory demanded an ideal situation, which never prevailed in nature. 'Yes, it would be a fine thing to put the world into an equation … Alas, how greatly must we abate our pretensions. The reality is beyond our reach when it is only a matter of following a grain of sand in its fall; and [despite all this] we would undertake the ascent of the river of life to its source!'

Yet, in such profound disagreement between the two great naturalists, there is also something touching and poignant. Fabre could hardly be dismissed as ignorant of biological science. What he represented was a different philosophical tradition harking back to an earlier age, a more innocent age, perhaps, with respect to its views of the divine nature of creation and the balances of nature. Scientific rivals Darwin and Fabre were destined to remain. But their disagreement did not translate to personal animosity or bitterness. In the words of Fabre's friend and biographer, Dr G.V. Legros:

> *It seems that on his side Fabre took a singular interest in the discussion [exchange of letters with Darwin] on account of the absolute sincerity, the obvious desire to arrive at the truth, and also the ardent interest in his own studies, of which Darwin's letters were full. He conceived a veritable affection for Darwin, and commenced to learn English, the better to*

> *understand him and to reply more precisely; and a discus-*
> *sion on such a subject between these two great minds, who*
> *were, apparently, adversaries, but who had conceived an*
> *infinite respect for one another, promised to be prodigiously*
> *interesting.*
>
> *Unhappily death was soon to put an end to it, and when*
> *the solitary of Down [Darwin] expired in 1882 the hermit*
> *of Sérignan saluted his great shade with real emotion. How*
> *many times have I heard him render homage to this illustri-*
> *ous memory![5]*

One can only sympathise with both men: with Darwin in thinking that Fabre had the genius and perception to open the wonder of insect metamorphosis to evolutionary understand-ing; and with Fabre, whose Homeric vision and religious awe at the complexity and beauty of nature rebelled against the very concept of evolution. Alas, posterity has confirmed that Fabre was as mistaken about Darwin as Keats was about Newton. Fabre was wrong in his interpretation of how natural selection works – natural selection in adapting an adult insect to a single source of food does not pretend to lead by any route to perfec-tion, only to signify adaptation to a specific life cycle, ecology, or sexual selection by the partner. Science, whether investiga-tive or evolutionary, does not erode our awe at the beauty of the natural world. On the contrary, as Fabre himself amply demonstrated, it enhances it.

In fact, Darwin's vision threw open a new window onto the understanding of life, the wonders of its origins, its beauty, and its diversity. And while the gentle Fabre refused to contribute his understanding of insects to this vision, there would be oth-ers, distinguished in their own right, prepared to take on that challenge.

Vincent Brian Wigglesworth was born into a medical family in Kirkham, a small town near Preston in Lancashire, on April

17, 1899, seventeen years after Darwin's death. In Provence at this time, Fabre was an energetic seventy-six, writing the books that would make him famous far beyond the field of entomology and never ceasing to 'multiply his pinpricks in the vast and luminous balloon of transformationism [evolution] in order to empty it and expose it in all of its inanity'. Both Fabre and Darwin would inspire Wigglesworth's lifelong love of insects, an inspiration that, however tortuous the road, and spanning almost the entirety of the twentieth century, would lead him to solve some of the mysteries of the butterfly's tale.

*Malaria mosquito*

9

# The Science of Life

ANYONE WHO HAS BEEN PRIVILEGED to observe young children in a garden cannot fail to notice their attraction to the miniature world of marching ants, buzzing bees, and creeping caterpillars, or the colourful flight of dragonflies, moths, and butterflies. Young Vincent was no exception. 'By the age of five', as he later recalled, 'I was keeping a large collection of caterpillars and other insects and spending hours and hours watching them'.

'In 1905, while caring for my cultures of *Abraxas grossulariata*, I discovered the metamorphosis of insects … A caterpillar I had imprisoned in a jam jar wrapped itself in silk and then, after a few days, emerged under my close and astonished observation, a butterfly!'[1] In fact it wasn't a butterfly, as the boy imagined, but a gorgeous moth, variously known in the British Isles as the magpie, currant or gooseberry moth. The caterpillars are a conspicuous pale green with bold black spots and a rusty line down the sides – a flare-like warning that they are likely to be distasteful to predators. Feeding on blackthorn, hawthorn, currant, and gooseberry bushes, the moth overwinters in caterpillar form. It pupates in May or June and metamorphoses, as Vincent had observed, in July and August to the striking adult moth with piebald wings of black and greyish-blue, with a yellowy-orange stripe in the middle of its forewings and this same flare of colour spreading, mantle-like, over the back of its head and shoulders.

It is hardly surprising that the emergence of such beauty from a chrysalis was such a wonder to the seven-year-old, who assumed he had discovered his very own fabulous secret. 'Of course I found in later years that this had already been known by Aristotle … Natural philosophers and small boys apparently find the same amazement in the transformation of insects from caterpillar to chrysalis to butterfly'.[2]

In the family garden was a willow tree whose leaves were invaded by red swellings, known as galls. He cut open the galls and was astonished to find that each gall contained a grub. It was another mystery he was determined to solve. But as he probed the galls he found no hole, or opening, through which the grub could have crawled into its tiny home within the leaf. 'I sought the guidance of my parents about this mystery – but in vain. From that moment I was a committed entomologist'.[3]

In 1911, the family moved from Lancashire to the south-east. 'For me it was like moving to El Dorado – so rich in insect life were the lanes and woods of Hertfordshire in those days'.[4] Wigglesworth's fascination continued to his school, Repton, in Derbyshire, where he devoted his leisure time to pursuing insects in the countryside. By now it seems altogether likely that Fabre was his role model, for he subsequently remarked: 'Although we had teaching in physics and chemistry, we had none in biology. In fact I would have felt it obscene to have these matters, about which I felt so deeply, treated as a school subject'. This, for a typical English gentleman, who would subsequently be described as 'gentle, reserved and formal, and with a wry sense of humour', is as close as we are going to get to his revealing a reverence as profound as Fabre's for his subject.[5]

In those early decades of the twentieth century, there were many patches of *harmas*-like wilderness in countryside around Repton where a boy might explore the world of nature. He was entranced by the immense variety of insects, which inhabit every nook and cranny of the biosphere, crawling over surfaces, burrowing through soil and decaying wood, flying through the air with a dexterity to match the most elegant birds, and even returning to recolonise the original womb of animal life, the world of fresh water and the oceans.

Insects may be only one class of arthropods on the tree of life, but it is a heavily populated class indeed: at least five-sixths of all animal species are insects. An estimate of the total number of insects on the planet runs to $10^{18}$. This is a million times a million times a million – too astronomical a number for the human brain to register other than in a conventional maths display, when it is 1,000,000,000,000,000,000.[6] It certainly puts our own human population total of $6.7 \times 10^9$ into perspective, though we constitute a single species. At a conservative estimate the insects that make up this single class may amount to some three million species.

During his time at Repton, Wigglesworth gave some thought to his future career. His father was a family doctor and he considered following the parental example. However, any such plans were interrupted by the Great War, when week by week at the school general assembly the headmaster would read aloud the names of pupils who had volunteered for action and were now counted in the ranks of the dead. And so, in due course, it came to Wigglesworth's turn, when he abandoned school, with just his secondary education completed, to become a signaller in the Royal Field Artillery. Fortunately, perhaps, he arrived late into the theatre of war and saw little real action. Yet his love of entomology enabled him to take comfort in beauty even amidst the carnage of the battlefields, as revealed by his subsequent observation that, in 1918, 'I was able to watch *Papilio macheaon* on the slopes of Vimy Ridge'.[7] One of the creatures that also delighted Fabre as a child, *Papilio machaeon* is the Old-World swallowtail, an attractive butterfly, banded black and maroon on bright yellow wings, with vivid red eye spots close to the elongated tail that gives the species its name.

After the war, Wigglesworth began his university education with preclinical studies at Caius College, Cambridge, where he completed a first-class tripos in biochemistry and physiology. By now he had affirmed his decision to forego clinical medicine – 'I felt I had already seen far too much of that in the home' – and set his mind on a future in medical research. His first posting was 'two very profitable years' working with the biochemist Gowland Hopkins, under whose direction he published eight scientific papers. Many of these were as assistant to a recent arrival from Oxford, the eminent J.B.S. Haldane, and involved Haldane as the experimental guinea pig in testing the effects of chemicals on himself. While conducting his biochemical research, Wigglesworth managed to find time for a diversion into his beloved entomology, studying and publishing papers on the pigments that coloured the wings of the cabbage white butterfly. In 1924, now equipped with his MD, he moved to St Thomas's Hospital in London, where he completed his MB medical qualification. Two years after Wigglesworth's move, Patrick Buxton was appointed head of the department of medical entomology at the newly established London School of Hygiene and Tropical Medicine.

Buxton was convinced that advances in infectious tropical diseases were being hampered by a lack of knowledge of the physiology of the insect carriers. He needed an entomologically trained lecturer to work with him and chose Wigglesworth on a provisional basis. Wigglesworth went straight to work. 'My first discovery, in an incubator in the laboratory, was the blood-sucking bug, *Rhodnius prolixus*, which had been brought to London from Venezuela by Professor E. Brumpt a year or two earlier'.[8] For Wigglesworth, this would prove a fateful happenstance. *Rhodnius* would be the subject on which he would conduct some of the most amazing experiments in the history of entomology.

But for the moment there were more pressing concerns for the young doctor-cum-entomologist. What he lacked, in practical terms, was experience of the tropics. Buxton suggested he travel to West Africa. So in 1927, Wigglesworth set out for Nigeria, where he spent six months travelling before moving on to the Gold Coast and Sierra Leone.

He soon had first-hand experience of African sleeping sickness, an epidemic infection caused by a single-celled organism known as a trypansosome, which is spread by the bite of the tsetse fly. Untreated African trypanosomiasis is frequently fatal. Malaria, the greatest of all the tropical killers, was also endemic throughout Nigeria, and bubonic plague was threatening to break out in the slums of Lagos.

Wigglesworth wrote regular letters home to a young lady, Catherine Semple, an aspiring artist whose vivacious and outgoing personality couldn't have been more opposite to his own. It would appear she had some reservations about the serious young scientist, wondering no doubt if life might be more fun with a less academically inclined suitor. But her father, Sir David Semple, himself a pathologist, would in due course instruct his daughter, 'Well, you know Wigglesworth is going to ask you to marry him. That young man is going places. If you don't accept him, I'm never going to speak to you again!'

In one letter home to his sweetheart, Wigglesworth described how he helped the local American-based team: 'I spent the morning with the Rockefeller yellow fever commission, while they performed autopsies on the front end of monkeys dead of yellow fever, meanwhile I collected lice from the hind end'. Yellow fever, a disease in which the yellow colour of the skin is caused by jaundice through massive liver destruction, is even more deadly than trypanosomiasis. It is transmitted by the bite of the mosquito and is one of the most lethal of the viral haemorrhagic fevers. In 1927 the viral cause had only just been discovered, and there was neither a cure nor a vaccine. When Wigglesworth arrived in Accra, in the Gold Coast, the main centre of yellow fever research, the local situation was tense. All infected monkeys had been destroyed and the laboratories were being decontaminated with cyanide vapours after two of the workers, the Japanese-American Hideyo Noguchi and the Irish microbiologist Adrian Stokes – the latter a joint discoverer of the viral cause of yellow fever – had only recently died from the fever they were researching. They had contracted the disease from a post-mortem on an infected monkey. Stokes's death in particular was a severe blow to tropical medicine, since this brave field worker had first discovered that *Rhesus* monkeys were susceptible to infection and, using that knowledge, had taken the first steps towards the preparation of the life-protecting vaccine that is now obligatory for travellers to the parts of Africa where the disease is endemic.[9]

Wigglesworth appeared to enjoy the experience of working in this dangerous maw, sucking up mosquito larvae out of crab holes and trekking through plague zones, accompanied by bearers carrying headloads of equipment, which included a self-contained laboratory complete with microscopes, not to mention his bed, bath, and furniture. During his travels, he discovered a number of unknown mosquito larvae and conducted original research on the physiology, digestion, and nutrition of the tsetse fly, which subsequently became the subjects of his first papers in medical entomology.

After his return to England, Wigglesworth was to spend many productive years as a lecturer and researcher at the School of Tropical Medicine and Hygiene, making an important contribution to our understanding of these diseases and their insect carriers. He took the opportunity to broaden his travels, studying insect-borne diseases in India, Burma, Malaysia, Java, and Sri Lanka, amassing a growing reputation for entomological research. In the midst of the Second World War, Wigglesworth was invited by William W.C. Topley, then secretary to the Agricultural Research Council, to help search for effective insecticides.

Wigglesworth was able to negotiate terms that would enable him to work on a broad and interesting canvas. 'We should be entitled to work on all aspects of insect physiology which might reasonably be considered to have a bearing on insect control', he said, 'and not only with insecticides'.[10] In effect he would be permitted to conduct "pure" research, since, as Wigglesworth saw it, all such research, in illuminating the life of insects, might in the fullness of time prove useful to agriculture – and also to medicine.

A year earlier, in 1939, the same year in which he was elected to the Royal Society, Wigglesworth published a landmark book, *Principles of Insect Physiology*, which would go through eight subsequent editions. The preface began with a statement that, in its universality of vision, was enlightening: 'Insects provide an ideal medium in which to study all the problems of physiology' – the science that looks at how living bodies work. No subject is more fundamental to life, bringing together biochemistry, molecular biology, heredity, reproduction, development, and even the wonders of evolution. For Wigglesworth, the way in which insects developed, from egg to adult, including metamorphosis, was primarily a question of physiology. His book, with its innovative breadth of thinking, would, in time, establish him as the 'father of modern insect physiology'.

At war's end, Wigglesworth was offered a position as head of a sub-department of entomology within the Department of Zoology at the University of Cambridge. He accepted the position and took the Agricultural Unit to Cambridge with him. His first preoccupation was ecological: he realised that DDT, then considered a wonder invention, might, in time, fail through the acquisition of insect resistance – an evolutionary phenomenon. He was also concerned that the intrusion of DDT into complex interactions between pests and their natural enemies might disturb the balances of nature. In the early post-war years, he was re-united in a social fellowship with his old college, Caius, and in 1952, he was elected to the Quick Professorship of Biology, a critical step to the freedom of research that had been his dream since boyhood – much as the freedom of the *harmas* had inspired Fabre.

In attempting to convey the importance of this professional liberation, I cannot improve on the words of his protégé, John S. Edwards, who, many years later, described his old professor as having a unique quality. 'It has been said of Alexander von Humboldt that he was the last person to know all of science ... Wigglesworth was probably the last person to know all of insect physiology'. So it was, in Edwards's words, 'Wigglesworth who brought together his own brilliant experiments with the work of others, from the 1930s to the 1950s, to present the first answer to the "how" question of post-embryonic development in insects'.[11]

'Post-embryonic development' being scientific jargon for the mystery of metamorphosis.

*Helicopter damselfly*

## 10

# Elementary Questions and Deductions

IN 1923, WIGGLESWORTH'S FIRST scientific paper had been published, a study of the effects of insulin on blood phosphate based on his work with Hopkins and Haldane at Oxford. Over the ensuing ten years at the London School of Hygiene and Tropical Medicine, a succession of articles display the progress of his interests, from purely medical to medical entomology – digestion in the tsetse fly, delayed metamorphosis in a predaceous insect, the function of the anal gills in mosquito larva, a theory of insect respiration. Then, abruptly, in 1933, like the broadside of a battleship erupting out of the tranquil dawn mists of a hitherto quiet ocean, the first of a classic series of papers heralded a new and all-consuming focus: an in-depth scrutiny of insect metamorphosis.[1]

Renowned for brevity in his communication, Wigglesworth often condensed his scientific writings to less than a page. But this opening paper, his first addressing the theme of metamorphosis, ran to some fifty pages.

It was clear that Wigglesworth knew he was confronting a very great challenge. But before he could even begin to investigate metamorphosis, he needed to take a hard look at what the mystery actually embraced. Rather like Sherlock Holmes, whose inventor, the contemporary Londoner Sir Arthur Conan Doyle, had died just three years before, Wigglesworth probed a seemingly intractable mystery using a series of elementary questions and deductions. His first question was obvious: why was it that all insects do not undergo metamorphosis?

What fossil evidence there was, proved both surprising and pertinent: most authorities had concluded that the earliest insects did not undergo metamorphosis. The most primitive insects are found as fossils in rocks from the early Devonian period, around four hundred million years ago, but none of these appears to be winged. The oldest evidence of winged insects appears to date to about eighty million years later, and these did not yet include the fully metamorphosing insects, such as butterflies, moths, bees, and beetles. This was a remarkable clue, for it implied that the more dramatic changes of insect metamorphosis had entered the insect life cycle long after the more primitive adult forms had already been defined. This made Wigglesworth's objective more elusive. Marine arthropods, judging from their fossils, predate the land insects by more than a hundred million years, and we know that, at some stage, these too had evolved a diverse array of metamorphic life cycles. But now it would appear that the metamorphic life cycle of the insects had not arrived, dripping with more than a hundred million years of marine ancestry, when those pioneering forms first emerged onto dry land. He had no option but to go back to first principles and consider the evolutionary history of the insects. When did the ancestral species crawl out of the oceans, and what do we actually know about their subsequent history?

What a strange and alien landscape would have greeted these earliest arthropod explorers of dry land, four hundred million years ago! There was none of the familiar vegetation that clothes the modern world in its dazzling variety of grasses, forests, and flowers. No birds flitted among the trees, no dinosaurs lumbered across the plains. It was the seas that teemed with complex life; on land there were no vertebrate animals of any description. Even the continents we recognise today did not exist. The entire landmass was joined up in a single supercontinent, Pangaea, with the green of the primitive ground-hugging plants, such as algae, liverworts, club mosses, and the stubby and knobbly Cooksonia, now extinct, struggling to gain a foothold around its inclement shores. It was into this harsh landscape that those ancestors of the insects first crawled out of the oceans, struggling merely to survive before further establishing themselves in damp shady places.

Wigglesworth focused on the arthropods, assuming from what was known of early marine arthropods that the hypothetical common ancestor of this gigantic and highly diverse group of invertebrate animals would have been a marine-based, creepy-crawly creature, with a body made of ring-like segments, each supported by a pair of appendages that, for the most part, functioned as legs. The creepy-crawly's front segments fused to become the head, and the head appendages became the mouth parts and antennae. The nerves within the head developed into a ganglion, or primitive brain, which collected the information arriving from its evolving eyes, antennae, and other sensory organs, and translated this into the appropriate movements and responses – attack, avoidance, devouring prey. Some of the appendages at the tail end became modified for mating – Wigglesworth had looked into aspects of this in his experiments in Africa; others served to manipulate the eggs as part of reproduction, or to carry sensory organs for various purposes.

The bodies of insects were too flimsy to leave much in the way of fossil evidence, but biologists thought the hypothetical insect ancestor most likely resembled a centipede, a few millimetres long. Inhabiting the shore lines, many of those early arthropods continued to evolve in oceans while others invaded the land, giving rise to the fantastic variety of creatures we see today, such as the crabs, lobsters, and shrimps in the oceans and the spiders, scorpions, and harvestmen on the land. The insects thrived among a subgroup of arthropods known as the mandibulates, which, as the name suggests, use side-mounted jaws to chew their food. So successful was their colonisation of land and air that the mandibulates still account for the largest number of species of any animal group on the tree of life, including forms that closely resemble their non-metamorphosing ancestors, the familiar millipedes and centipedes.

In his paper, Wigglesworth had arrived at an elementary deduction, one that derived from this common arthropod history and which might provide him with a next clue to solving his mystery: '[The process of moulting] is the homologue of metamorphosis, and there can be small prospect of understanding the complex process of metamorphosis until the physiology of [moulting] has been adequately described'. Even those insects that did not metamorphose went through the process of moulting. Very well – here was where he would begin!

As he took a careful look at this common characteristic of the arthropods, it surprised him that previous investigators had devoted so little attention to the changes that enabled moulting in insects, whether or not they underwent metamorphosis. As an animal needed to grow, much like other animals, the insect was faced with an unusual dilemma. It did not have an internal skeleton, as did a mouse or a human. Its skeleton was on the outside – an exoskeleton – and was composed of an extremely rigid structure, the cuticle, which was produced by the insect skin or epidermis. The problem facing the insect was simple: how to grow, safely and effectively, despite the restrictions of the rigid exoskeleton.

Wigglesworth looked at what was known of the remarkable insect epidermis and its role in the evolution and physiology of insect life.[2] It was obvious that, in pioneering their way onto dry land, the ancestors of the insects would have needed both to conserve water and to protect their vulnerable bodies with defensive armour. They had evolved the ability to convert the carbohydrate found in mucoproteins (like the slime of slugs) into a tough fibrous secretion called chitin. Chitin is a polymer, rather like the cellulose in blotting paper. It comprises long chains of chemicals linked together to form a cohesive layer, which forms the basis of the surface covering of all insects. Over most of the insect body the chitin is toughened to make the stiff horny exoskeleton through the addition of a protein, called sclerotin, which, when it becomes tanned in sunlight, makes it even tougher, like the shiny black carapaces of beetles. Sclerotin is also the ideal material for making the thin membranes of insect wings, or the needle-like piercing stylets of mosquitoes and the stings of wasps and bees – even the elaborate mating organs of the two sexes. The durability of this substance is made possible by chemical cross-linking sulphur groups, rather like, but much more resistant than, plastic. A similar cross-linkage of sulphur groups is involved in the formation of the scales of reptiles, the feathers of birds, and the hair, nails, hoofs, and horns of mammals. The carapace of lobsters is hardened in a similar way through the addition of lime to the cuticle, though in the lobster the harder parts, that is the tips of the mandibles and the claws, are also coated with sclerotin.

Anybody who has tried to kill a cockroach will know how tough and protective this surface armour is. It is also resistant to many harmful chemicals, even to natural decay, as archaeologists realise when they employ the identity of insects in soil or strata to deduce the ecology of dig sites that are thousands, or tens of thousands, of years old. Moreover, this surface armour has an additional, extremely useful property. Kept supple, for example in the region of joints, it allows the flexion necessary for movement. According to Wigglesworth, 'Just as the invention of keratin made possible the flying equipment of the pterodactyls, birds and bats, and the hairy hugginess of mammals, sclerotin made possible the extraordinary evolutionary diversity of the insects ... Indeed, it is impossible to study the natural history of insects without discovering at every point how dependent they are upon this remarkable plastic'.[3]

We mammals also enjoy great flexibility in our bodies and limbs, but we do so through a very different technology – an internal skeleton, equipped with a wide variety of joints, all surrounded by stretchable, soft flesh. But our internalized skeleton also makes us vulnerable to wounding of our skin and organ tissues. The larval forms of butterflies are equally vulnerable, thanks to a highly folded and soft skin, which is necessary to allow the caterpillar to grow at incredible speed. Similarly, the belly of the queen termite is so elastic that, as her reproductive organs fill up with eggs, it can expand fifty-fold. The adult insect, on the other hand, dons its suit of armour to face the hostile world, allowing only select modifications to support the flexing of joints and powerful jaws.

While the insect is better defended than the mammal, that same armour has one overwhelming disadvantage: a rigid external skeleton makes further growth impossible. Unlike vertebrates, such as fish or mammals, whose internal skeleton can readily grow during childhood to attain the full adult size, insects – and arthropods in general – can only grow if they first discard their outer shells. This is why insects, crabs, and lobsters moult.

It is important to distinguish moulting from the more complex concept of metamorphosis: moulting merely permits an increase in size; metamorphosis implies a change in form.

More primitive orders of insects keep to a primitive blueprint: they moult but they do not metamorphose. These include the silverfish, which flitter across the floor from under their cozy rugs at night; the springtails, which use a forked appendage on their abdomens to leap through the air; and a number of less familiar forms that live in the humid mats under vegetation or the moist litter of forests, skulking under logs or stones, environments similar to the primal ecology of the terrestrial pioneers.

It was from careful observation of these early forms that Wigglesworth was able to make another elementary deduction.

All primitive insects are wingless. This, taken with other considerations, suggested that insect metamorphosis might be linked to the invention of flight.

If there were seven wonders of the biological world, the flight of insects would be one of them. Ornithologists and entomologists might debate, grace for grace, the flight of a swallow with that of a butterfly, yet insect flight, with its extraordinary range of strategies and dexterities – the ability to take off backwards, to hover and soar, or to alight upside down on a ceiling – is arguably more complex and inventive than that of the birds. The imagined flight of fairies, including Tinker Bell in the Peter Pan stories, more closely resembles that of an insect than any bird, much as David Attenborough in his *Life in the Hedgerow* series so charmingly demonstrated with the helicopter damselfly, the largest damselfly in the world, with wings of sapphire that extend to twenty centimetres and exhibit a balletic elegance in motion. Yet such is its heart and purpose that, in the courtship of the cascade damselfly – a rare species limited to a small number of Central American waterfalls – the male, which weighs no more than a sprite, will endeavour to impress his beloved by flying through the falling torrent.

The evolution of flight is surely one of the main reasons for the extraordinary success of the insects, helping their hunt for food, their escape from danger, their search for a mate, and their exploration, far and wide, of new habitats to make their own. In 2003 Michael Engel, an evolutionary biologist at the University of Kansas at Lawrence, was researching background information for a book on insects. Needing to describe a fossil arthropod, he peered into the contents of a dusty drawer in the bowels of the Natural History Museum in London and gasped at what he discovered. Fossilised in chert rock that had been dug up near Aberdeen in the 1920s, were the remains of an extinct insect, *Rhyniognatha hirsti*, which must have been one of the earliest insects ever found. Though the fossil did not include wings, Engel and his colleague, David Grimaldi, found that *Rhyniognatha*, which was about five millimetres long, had jaw articulations commonly associated with modern-day winged insects.[4] This might document a stage in insect evolution prior to the development of wings or represent the first step in metamorphosis among insects. The fossil was estimated to date from 390 to 408 million years ago – an age that, if winged *Rhyniognatha* were to be confirmed in other examples of the fossil record, would push back the date for the evolution of true insects to more than four hundred million years ago. It would also push back the evolution of metamorphosis to an earlier era than had previously been assumed, perhaps surprisingly early in the history of the insects.

Dragonflies do not go through the stages of larva, pupa, and adult. But they do undergo a less dramatic change, particularly in the final moult, known as "incomplete metamorphosis". Dragonflies and damselflies belong to the order of insects known as Odonata, which counts some five thousand species, including the giant tropical variety, *Megaloprepus caerulatus*, the helicopter damselfly. This sizeable insect measures up to nineteen centimetres from wingtip to wingtip and yet, thanks to its hovering ability, can sneak up on spiders at the very heart of their webs and devour the succulent abdomens while the unfortunate victims are still alive. These are the closest living relatives of the almost science-fictional *Megasecoptera* that, judging from their fossilised remains, thrived in the swamps and forests of the Carboniferous era, more than three hundred million years ago. Gigantic creatures, with a wingspan of almost a metre, *Megasecoptera* resembled vampires with their grotesque sucking mouth parts. Their pond-dwelling larvae were ferocious predators; they grew to thirty centimetres long.

Fossilised *Megasecoptera* had been known since 1885, and their extraordinary forms would have been familiar to Wigglesworth. Equally familiar to him was the fact that dragonflies have been remarkably resistant to change during their long evolutionary existence, many still holding their wings outstretched at rest. (The more graceful damselflies have learned how to fold their wings together above their backs.) Though mere dwarfs in comparison to their ancestors, the larvae of most dragonflies still grow to about five centimetres long. They are aquatic and come equipped with gills to breathe under water, big eyes to help them spot their prey of insects and tadpoles, and an extensible lower jaw that shoots out and hooks their victims. And here, in the development of this ancient flying order we find another elementary clue to the evolution of metamorphosis. The dragonflies develop their wings on the outer surface of their bodies, from buds that are present during the larval stages and that grow at intervals throughout a series of moults.

Fossil nymphs of these ancient insects show scaled-down "winglets" that appear to have been articulated but held out stiffly to the side. These subsequently evolved to immovable pads folded over the back, protecting the developing wings from harm. This in turn ushered in the first incomplete form of metamorphosis, with the external pads we still see today in the larvae of certain insects. Complete metamorphosis, such as we see in butterflies and bees, evolved later, yet still more than three hundred million years ago. Some entomologists believe there was a big expansion in the more colourful examples of complete metamorphosis, including among moths and butterflies, at roughly the same time that the flowering plants appear in the fossil record, suggesting the possibility of an interlinked evolution, known as co-evolution, which wrought some of the loveliest jewels of the plant and insect worlds.

What, Wigglesworth now asked himself, was the real difference between incomplete metamorphosis, such as we find in the dragonflies, and true or complete metamorphosis, such as we find in butterflies?

Of the twenty-eight orders that make up the Insecta class, the four most primitive orders moult but do not metamorphose and fifteen other orders undergo incomplete metamorphosis, the wings developing from buds on the exterior of the cuticle, with the buds gradually increasing in size with successive moults to complete their development in a final growth spurt. Besides the dragonflies, these external-bud orders include the mayflies, cockroaches, mantids, termites, earwigs, and stick insects, as well as the grasshoppers, locusts, and crickets. In all of these, we observe a process of gradual growth through moults as the insect develops from larva to adult, albeit the greatest change takes place in the last moult. In these incompletely metamorphosing orders, there is no separate larval phase, and there is no pupal stage before the final metamorphosis.

By contrast, in the remaining nine orders, which undergo complete metamorphosis, the wings develop internally. The more specialised of these orders include the translucent-winged wasps, bees, and parasitic ichneumon flies; the double-winged true flies; the hundreds of thousands of species of beetles; and the quarter of a million species of butterflies and moths. Among these last, the most varied and colourful of the insects, there is always a larval stage, such as the caterpillar or the grub, often accompanied by voracious feeding and growth. Most telling of all, the transformation involves the bizarre process familiar from those earliest lessons at school, the same mystery that has beguiled naturalists since Aristotle, in which the caterpillar stops feeding and becomes cocooned from the world in a silky or leathery case, called the pupa or chrysalis, within which extraordinary changes in form and development take place, to result in the emergence of the fully-formed adult insect.

*Goliath beetle*

## 11

# The Phoenix in Its Crucible

WHEN I VISITED Jonathan Wigglesworth, Vincent's son and himself Emeritus Professor of Perinatal Pathology at Hammersmith Hospital in London, he showed me a surviving display case from his father's personal collection. I found myself gazing at an array of insects laid out like jewels, glittering and sheening, the majority of which I could never have put names to, yet all perfect in their sculptured forms. Prominent among them was a goliath beetle he might have brought home from his African safari. I couldn't help but wonder if, in that fabulous but slightly monstrous creature, he had considered the mystery of its birth through that striking transformation that had entranced him as a boy.

The African goliath beetle is the heaviest flying insect in the world. It belongs to the same family as the scarab, which was holy to the priests of ancient Egypt, and it can grow to twelve centimetres long and weigh up to a hundred grams. The female beetle lays her eggs, which are no more than a few millimetres in diameter, on the rich, moist, rotting detritus of the tropical forest floor. When the creamy caterpillar-like larvae hatch, they feed on dead wood, but as they grow bigger and stronger they become predators of other insects. This carnivorous diet is quite nutritious, enabling them to grow so rapidly that within a few months they reach fifteen centimetres in length. At this stage, they burrow into the forest floor and construct a protective cocoon. Here they pupate, undergoing a complete metamorphosis, before the gigantic adult beetle emerges fully grown, with its black-and-white-striped thorax and its first pair of wings, evolved to hardened shields called elytra, which protect the flight wings – an apparatus so flexible and delicate, and yet so powerful, that when the goliath takes to the wing it sounds like a miniature helicopter.

This extraordinary transformation within the pupa is one the goliath shares with the honey bee, the monarch butterfly, and Fabre's giant peacock moth – as well as with the entire order of beetles, Coleoptera, with its 275,000 members. The larva of the goliath beetle is obviously a very different animal from the adult, much as the larval grub is to the honey bee, wasp, or blue bottle fly, or the caterpillar to the butterfly or moth. For the larva to metamorphose to the giant, helicopter-buzzing adult, massive structural change must take place throughout the length, breadth, and internal structures of the larval body. Thus, rather than a den of repose, the enclosed chamber of the goliath's pupa really is a crucible tantamount to the mythic pyre of the phoenix, where the organic being is broken down into its primordial elements before being born again. The immolation is not through flame but a voracious chemical digestion, yet the end result is much the same: the emergence of the new being, equipped with complex wings, multi-faceted compound eyes, and the many other changes necessary for its very different lifestyle and purpose.

The new adult needs an elaborate musculature to power its wings. These muscles must be created anew since they are unlike any seen in the larva, and they demand a new respiratory system – in effect, new lungs – to fuel them, with new breathing tubes, or tracheae, to feed their massive oxygen needs. These same high energy needs are supported by changes in the structure of the heart, with a new nervous supply to drive the adult circulation and a new blood to make that circulation work. We only have to consider the dramatic difference in appearance between a feeding grub or caterpillar and a flying butterfly or a beetle to grasp that the old mouth is rendered useless and must be replaced with new mouth parts, new salivary glands, new gut, new rectum. New legs must replace the creepy-crawly locomotion of the grub or caterpillar, and all must be clothed in a complex new skin, which in turn will manufacture the tough exoskeleton of the adult. Nowhere is the challenge of the new more demanding than in the nervous system – where a new brain is constructed de novo. And no change is more practical to the new life form than the newly constructed genitals necessary for the most essential role of the new adult form – the sexual reproduction of a next generation. The overwhelming destruction and reconstruction extends to the cellular level, where the larval tissues and organs are broken up and dissolved into an autodigested mush in which individual larval cells follow new developmental instructions and, to all intents and purposes, return to the embryonic state, with the constituent cells becoming undifferentiated in form.

This process is as startling in its evolutionary implications as for us today as for Vincent Wigglesworth three quarters of a century ago. In Wigglesworth's own words, 'These remarkable transformations raise questions which have exercised the minds of naturalists from the earliest times'. And so he asked the pertinent questions in his deductive exercise: 'How are the changes in form regulated? How did they arise in the course of evolution? What is the significance of the pupal stage?"[1]

In the 1930s, when Wigglesworth first began to explore metamorphosis, many of his colleagues believed the answer to all three questions lay in structures known as imaginal discs. The term "imaginal" had nothing to do with the imagination, but with the root word from which imagination itself springs: the Latin *imago*, meaning an image or likeness. The imaginal discs are curious thickenings of the inner surface of the larval skin, or ectoderm. First described in 1762 by the Dutch naturalist and engraver Pieter Lyonet, whose monograph on the anatomy of the goat moth is regarded as one of the most beautifully illustrated works on anatomy ever published, imaginal discs were later found to occur only in insects that undergo full metamorphosis. Through careful study, biologists discovered that, during the pupal stage, the imaginal discs grow and transform into the complex structures of the adult, including the legs, the wings, the genital apparatus of male and female – even the striking multi-faceted compound eye. In 1864, the German naturalist August Weismann proposed that the imaginal discs were actually nests of embryonic cells – what we, today, would call stem cells – containing the genetic blueprint for the future image of the adult insect. These lie dormant through the life of the larva only to be activated during the furious demolition and reconstruction within the pupa, when, on some critical inner signal, they grow and develop into the specialised organs and tissues of the adult insect.

Wigglesworth had reservations about the purportedly unique role of the imaginal discs in metamorphosis. Interpretations of metamorphosis at the time centred around the more extreme examples, but, as he said, 'many parts of the adult body, even in such insects as Lepidoptera [moths and butterflies] or Hymenoptera [wasps, bees, and ants], which suffer a spectacular metamorphosis, are not formed from imaginal discs but are laid down by the same cells as have formed the larva'. Moreover, even in the completely metamorphosing insects, there were exceptions to the wholesale immolation in the furnace of the pupa. For example, in some insects, such as the alderflies and scorpionflies, as well as several of the less developed moths and butterflies, some larval muscles were retained to help create the abdomen of the adult. Even the superficial inertia and rigidity of the pupa was seen to vary. And in the pupa, the imaginal discs were not the sole source of adult tissues.

Such observations implied that developmental mechanisms other than the imaginal discs were at work in the pupa – those same, or very similar, mechanisms that were at work in the less dramatic changes seen in the orders that underwent an incomplete pattern of metamorphosis.[2] All of this suggested to Wigglesworth that perhaps the dramatic changes seen in the pupa of butterflies and moths might not be as representative as the proponents of imaginal discs assumed.

Wigglesworth accepted, with reservation, the Darwinian viewpoint that the larva is subject to the influences of its environment. Through 'natural selection (or whatever other agencies control organic change) it may undergo an evolution of its characters that leave the adult insect untouched', he wrote.[3] It made perfect sense to Wigglesworth that natural selection was operating independently on the larval and adult stages, living as they did in their very different life cycles. It seemed a predictable sequel to this that as the differences between larvae and adults became more extreme, for example between the caterpillar and the butterfly or the maggot and the fly, evolution would need to find some adaptive solution to the increasingly wide gap in body forms and physiology between the larval and adult insect. Here, to his mind, lay an important clue.

What if, putting the curious mystery of the pupa to one side, the key to metamorphosis was not confined to butterflies and moths, but extended to all orders of insects other than the primitive non-flying ones, whether their metamorphosis was incomplete or complete? To put it bluntly, what if there was no significant difference in the mechanisms of incomplete and complete metamorphosis?

If he was right on this crucial assumption, he could direct his research towards finding a single explanation that would satisfy all.

Aristotle taught that the embryonic life of insects merely continued until the formation of the perfect adult. We might recall his words, 'The larva while it is yet in growth is a soft egg'. Wigglesworth disagreed. He took the view that the larva, pupa, and adult were a series of alternative forms assumed by a complex organism at different periods in its development. The fundamental factor, in Wigglesworth's mind, was what exactly determined the form it assumed at any one stage.

He began to devote more and more of his time to his study of the great mystery. His day-to-day experimental methods are not revealed in his scientific publications, but it is possible to extrapolate somewhat from the recollections of those who worked with him.

Henry Bennet-Clark, today Emeritus Reader in Zoology at the University of Oxford, was from 1957 until 1960 a graduate student working under Wigglesworth's supervision at Cambridge, where he studied moulting in ticks and spiders before moving on to the mechanical properties of the expansile skin, or cuticle, an interest that would lead him to become a world expert in modelling insect movement.[4] In his years at Cambridge, he had plenty of opportunity to observe how Wigglesworth thought and worked. His first, and most telling, observation was that Wigglesworth almost always worked alone. 'This, I think, is reflected in his publications, which were something in the region of 250, of which I think only seventeen had more than one author'. In this he closely resembled another medical graduate and famous scientist, Jorgen Lehmann, who discovered the anti-tuberculosis drug, para-amino salicylic acid, or PAS.[5]

'Why do you think he was so solitary in this?'

Suddenly our interview sprang alive. 'Wigglesworth once said to me, "Okay – we'd better arrange for you to get hold of some ticks." And duly, some ticks appeared. I, now, thought – "Fine!" I was happy with this. Then he said, "You'd better have a rabbit to feed them on." I duly got a rabbit in the animal house. I bred up my ticks and started doing my experiments. After about six weeks, he said, "Well – how's it going?" By then I'd learnt how to grow ticks. I'd started finding out some odd things about their biology, which put me onto the line of research I subsequently followed. I suppose he knew that I was there. He would come in, say at quarter to nine every morning, and go out at quarter to one. Lunch at Caius. Come back at two. Go home again at about five. He had his bicycle outside the lab somewhere. [Wigglesworth never learned to drive]. He cycled off to the northern part of Cambridge, where he lived. His daily routine was like that. I have no evidence whatsoever of what he did before a quarter to nine, nor what he did after five. I suspect he, as it were, wrote *The Principles*, and so on. He was not what might be called an outstandingly clubbable man.'

'But you must have talked to him about your work?'

'Every so often, when I thought I had got hold of something different – or any of us grad students thought we had something different – we'd go and see him and he'd sort of grunt.'

'Was this in his office?'

'Either in his office or in his lab, where he would be bent over whatever he was doing to *Rhodnius*. And he would grunt. Every so often, when you'd actually done something good, he'd say, "That's quite interesting." He would sometimes say, "Yes. I think that might be worth pursuing."'

Other characteristics, even more revealing than the air of pre-occupied scientist, ran through all of Wigglesworth's experiments. Bennet-Clark captures them succinctly: 'First of all [his work], and particularly so his early papers, were immaculately illustrated. I think he must almost have thought as he was drawing: "If I do this properly, it's going to be in books." And they jolly well were, of course.'

Another important characteristic of Wigglesworth's work was the small number of experimental subjects. As Bennet-Clark explained, 'If you construct an experiment to test a prediction and that prediction is borne out by results on five separate occasions, you can say, "Whoopee! That is a one in thirty chance. That's a three percent probability. And I don't need to do it again." If you read some of his papers you will actually find that it is quite obvious that the sample size was three. Wigglesworth, on several occasions, only did something once.'[6]

At the same time Bennet-Clark was working on the biomechanics of the insect cuticle under Wigglesworth, Simon Maddrell was working on urine production. Maddrell, who subsequently discovered the insect diuretic hormone, is now a Fellow of the Royal Society, and holds the same professorship once occupied by Wigglesworth at Cambridge. He also recognised the unusual clarity of Wigglesworth's experimental thinking.[7] Such small sample sizes were not the mark of carelessness or laziness, he said. On the contrary, they implied deep thought and very careful planning of experiments in advance. In essence, Wigglesworth formulated an experimental approach in which he sought a specific answer: yes or no. The experiment was constructed with such elegant simplicity and lucidity that very few test subjects were necessary.

Wigglesworth applied the same clarity of approach when he perceived that the physiological factors controlling metamorphosis could be encapsulated in two simple questions: What factors inhibit the development of the imaginal discs in the young stages? And what factors cause the larval tissues to disintegrate and activate the imaginal buds at metamorphosis?

If he was right in ignoring the widely held views of critical differences between incomplete and complete metamorphosis, then the answers to these two questions would prove to be the same across all examples of metamorphosis. He could assume the latency and subsequent realisation of the adult insect can be understood just as well from incompletely metamorphosing insects, in which imaginal discs do not occur. Conveniently, he was already familiar with an insect that fitted the bill, an experimental subject that was tough and hardy, easy to keep alive in the vaulted underground bunkers, known as insectaries, of the London School of Hygiene and Tropical Medicine – an insect he had already found amenable to clear and simple experiment.

As Wigglesworth himself recalled: 'It was on these grounds that *Rhodnius prolixus* was chosen for the study of metamorphosis'.

*Rhodnius prolixus*

## 12

# *Two Souls in One Body*

*RHODNIUS PROLIXUS*, OTHERWISE known as the "kissing bug", is a parasitic insect of the drier savannah areas extending from southern Mexico to northern South America. It is a member of the order Hemiptera, commonly referred to as "bugs", many of which, as Wigglesworth himself expressed it, 'interfere seriously in the affairs of man'. The Hemiptera include more than fifty-five thousand species, most of which, such as cicadas, aphids, and greenflies, have penetrating mouth parts that enable them to feed on the sap of plants. Others use those same syringe-like mouths to suck the body fluids of other insects or, like *Rhodnius*, to suck the blood of mammals and birds.

During its life cycle the wings develop outside the body, progressively enlarging during successive moults from externally placed lobes or pads. There is no dramatically different larval phase, such as we see in the caterpillars of butterflies and moths, and no pupal stage. Instead the insect grows and develops through a series of moults, completing growth and development during the final moult. In this way *Rhodnius* is a typical example of incomplete metamorphosis. Sap- and blood-sucking insects are notorious for transmitting disease-causing microbes, particularly viruses, from plant to plant or from animal to animal, and *Rhodnius prolixus* is no exception. Fairly big for a disease-transmitting insect, the fully grown *Rhodnius* is about three centimetres long, a fact I was privileged to witness when a medical entomologist, Professor Chris Curtis, let me handle living specimens in the underground insectaries at the London School of Hygiene and Tropical Medicine. Its body is shaped like an extended prism, with a long, narrow cylindrical head and an abdomen capable of enormous syringe-like expansion, an adaptation designed to accommodate its predatory lifestyle. The larval stages of *Rhodnius* ingest from six to twelve times their own weight of blood at a single meal in preparation for moulting. In explaining why he chose this insect for his study of metamorphosis, and for many subsequent experiments, Wigglesworth freely declared: 'It so happens that in *Rhodnius*, this single meal of blood suffices for each [larval stage], and that moulting occurs at a definite interval after this meal, the whole process following a more or less constant timetable. This circumstance makes *Rhodnius* an admirable subject for such an investigation'.[1]

In its natural habitat, *Rhodnius* roams across a number of ecologies, including the crowns of palm trees, from where it swoops to feed on the blood of marsupials, rats, reptiles, carrion birds, and humans. It is the insect vector of the parasite *Trypanosoma cruzi*, which causes an extremely unpleasant and largely incurable illness, Chagas disease, a condition afflicting the intestines, heart, and nervous system. Some authorities suspect this was the real nature of the illness that plagued Charles Darwin for decades of his life and that eventually caused his death.[2]

This insect now took central stage in Wigglesworth's intensive exploration of metamorphosis. And like Aristotle, Darwin, and Fabre – indeed, like every great naturalists that had ever gone before him – he began by making full use of intelligent observation.

The larval stages of an incompletely metamorphosing insect are known as nymphs, and the moulting from one larval stage to another is called ecdysis. *Rhodnius* moults through five nymphal stages, during which the structures and pigment patterns of the surface cuticle show little change. There is progressive increase in size and development of the wing lobes and, in the later stages, the genitalia begin to appear. But these changes are slight compared to the major development that occurs in the final moult, when the fifth nymph changes into the adult. The structure and colouring pattern of the cuticle is dramatically transformed. Compound multi-faceted eyes now appear for the first time. Elaborate genitals are fully developed. Meanwhile, the thorax, or chest, adopts its final size and shape, strengthened to accommodate the flight muscles, and the wing lobes are massively transformed into the large and powerful wings.

For Wigglesworth, who was convinced that incomplete and complete metamorphosis could be considered as essentially the same mystery, 'It is therefore convenient ... to refer to this final stage as "metamorphosis"'.[3] Now, examining the changes that took place during this metamorphosis, it was clear these changes embraced the entire body of the insect – skin, eyes, sexual organs, muscles, and wings. For such universal changes to happen abruptly, during the final moult of an insect such as *Rhodnius*, there had to be some form of a message, some signal, capable of being transmitted to every tissue and organ throughout the insect body. Very few systems were capable of body-wide transmission in this way. Wigglesworth already knew, from the work of other biologists, that the wing muscles required an intact central nervous system for their momentous development during the final moult. And the nervous system was, of course, capable of transmitting a message or signal, body-wide. Now he asked the next obvious question: did the nervous system hold the secret to metamorphosis?

The European gypsy moth, *Lymantria dispar*, is a member of the tussock family and a considerable pest in forestry. The adult female is pretty, if decidedly hairy, and its creamy wings, dotted with dark spots, extend to about five centimetres in span. It is too large and heavy to fly; instead, it hauls itself up onto the trunks of trees and shrubs, where it exudes a pheromone to attract the smaller and browner male. Once mated, it lays a single egg mass, after which the female dies; up to a thousand caterpillars emerge from the eggs at the end of April or in early May. These caterpillars are both ravenous and omnivorous in their tastes. Such is their hunger, they feed on more than 450 different species of plants, but their preferred diet consists of broad-leaf oaks and alders and needle-leaf Douglas Fir and Western Hemlock. They are capable of completely defoliating trees. When an infestation is extensive, the result, in arboreal terms, is an epidemic.

In 1923, Stefan Kopeč, Professor of Biology at Warsaw University, chose the gypsy moth as the subject of a series of experiments aimed at uncovering the role of the nervous system in metamorphosis. In one experiment, he removed the nerve centres, known as the thoracic ganglia, from pupae and waited to see how this surgery would affect their subsequent metamorphosis. The wing muscles failed to develop. But this was not critical, as Kopeč went further to show that growth of the body form of the adult, which is derived from the epidermis, was otherwise normal.[4] Another scientist found that if you removed the corresponding nerve ganglia in the praying mantises, *Sphodromantis*, the antennae and legs still grew to normal size and shape – in spite of the fact that they contain no nerves or muscles. Wigglesworth got the message – the nervous system was not the ultimate key to metamorphosis.

He switched his focus to the skin. Human skin is a complex organ, but it does not determine our overall shape. Like all vertebrates, our inner skeleton performs this function, not only controlling our size and body form but also affording fulcrums for our muscles and armour for our vulnerable organs, such as the brain and heart. The insect epidermis fulfils these numerous roles. And different areas of skin in the same *adult* insect will exhibit different appearances and properties.

When Wigglesworth excised a small area of the skin of a fourth-stage nymph of *Rhodnius* and implanted it into a *different area* of the body surface on another fourth-stage nymph, the transplanted skin healed into its new site and looked similar to the rest of the skin in that area. But when the recipient nymph metamorphosed to an adult, the implanted skin developed the adult pattern and structure of the part of the body from which it had been taken. Through this experiment and others, he proved that even in insects that undergo incomplete metamorphosis, major changes in epidermal structure occurred as part of the development from nymph to adult.

He sat back and thought about it all again. It seemed evident the fertilised egg contained the hereditary programming for two very different and specialised patterns of body development, one larval and one adult. But the expression of the adult programme was intrinsically different in the two major forms of metamorphosis. In complete metamorphosis the adult features arose from stem cells that had been set aside in the imaginal discs for the pupal stage; in incomplete metamorphosis, as in the case of *Rhodnius*, the adult body form was present yet latent throughout the larval cells at the time when they were expressing the nymphal pattern of development.

Some mysterious organ in the insect body must control this

curious pattern of duplex development. What organ was it? And how did it control development?

Wigglesworth reminded himself that metamorphosis from the insect larva to the adult, whether from nymph directly to adult or from caterpillar through the intermediate stage of pupa to adult, affected every cell in the insect's body at one and the same time. Given the nervous system had been excluded as the controlling influence, the only other medium capable of broadcasting a signal in such a universal manner was the blood stream.

In 1922, Kopeč appeared to have confirmed this in the development of the gypsy moth.[5] In Wigglesworth's view, the most likely explanation of how the blood stream would carry such a universal message was through a chemical signal – what today we call a hormone.

*Parabiotically conjoined insects*

## 13

# Bizarre Extrapolations

GOLIATH, IN THE BOOK OF DAVID, stood six cubits and a span – in modern terms nine feet, nine inches tall – while Tom Thumb, in traditional folklore, was said to be small enough to hitch a ride on a butterfly. In actual recorded history, the tallest man was Robert Pershing Wadlow, a native of Alton, Illinois, who was just over eight feet, eleven inches, while the tallest woman was Zeng Jinlian, who was born in China and grew to be almost eight feet, two. Up to the end of the nineteenth century, we had no means of knowing how such extremes of stature came to be. Yet in the lives of ordinary men and women, sudden growth, as well as wholesale changes in body size, body shape, and internal development are common experiences of the strange, often disquieting events of puberty.

Before puberty, boys and girls have the same proportion of muscle, skeletal mass and body fat. However, during puberty, boys and girls undergo dramatic physical development, including rapid body growth and major changes in muscle and fat distribution, which differ between the sexes. So profound are these changes that, by the end of puberty, healthy males have one-and-a-half times the proportional skeletal and muscle mass of females, whereas healthy females possess twice as much proportional body fat as males. These are accompanied by major cellular and tissue changes involving sexual and related organs, such as the development of breasts and genital organs in females and of genital organs in males. Today, we know the maturation process involves activation of developmental genetic pathways – the same sort of genetic pathways that are first involved during the embryo's development in the womb, so that, in the scientific jargon, human puberty is considered to be a post-embryonic developmental phase. More interesting still, recent research has ascertained that the processes of metamorphosis and puberty have many commonalities, so much so that some developmental biologists now suggest that 'puberty may be considered a variation of the metamorphic theme'.[1]

One of these commonalities is the involvement of hormones, tiny secretions that circulate internally from specialist endocrine glands and play a pivotal role in normal body development and health. With the discovery of human growth hormones in the 1920s and 1930s, "hormones" had become a buzzword in medical and mammalian animal studies.[2] For example, the thyroid hormone, secreted by the thyroid gland in mammals, was found to control the maturation and speed of the body's physiology and chemistry. Scientists also realised that puberty was brought about in humans by pulses of a so-called gonadotropin-releasing hormone, produced by the part of the midbrain called the hypothalamus, which is exquisitely responsive to external cues and is involved in metabolism and other autonomic nervous functions. This hormone stimulates the master endocrine gland, the pituitary, to increase the secretion and

release of the sex gland-stimulating hormones, or gonadotropins, which travel through the blood stream to the ovaries and
testes, where they provoke increasing blood levels of oestrogens
and androgens. These hormones determine whether the body
develops as male or female. To the medically qualified Wigglesworth, all of this would have been very familiar.

Yet, there was a major problem in extrapolating the vertebrate hormone discoveries to insect metamorphosis: earlier
experiments appeared to prove that insects do not possess hormones. In Poland, Stefan Kopeč had suggested that his results
involving the bound gypsy-moth caterpillars might indeed indicate the presence of a hormonal signal, but the general opinion
held, and Kopeč's proposal had been largely ignored.[3]

Wigglesworth was aware that Kopeč had shown that the
insect brain was the source of the initial stimulus for moulting. When Kopeč beheaded gypsy-moth caterpillars, the brainless larvae didn't necessarily die, but the subsequent moult to
the pupa failed. Another colleague had demonstrated that if a
severed brain was re-implanted into the headless caterpillar of
the elephant hawk moth, *Deilephila elpenor*, it restored the ability to pupate. Much the same had been seen in the caterpillar
of the agricultural pests of the genus *Ephestia* and the silkworm
moth *Bombyx mori*. What then was the real nature of the metamorphosis control factor? Wigglesworth was increasingly convinced Kopeč had been right in suggesting it was a hormone.
And, if it truly was a hormone, the central question was: Where
was it being manufactured and released?

Wigglesworth explored this question in a series of bizarre
experiments now considered to be classics in the study of insect
physiology. In his preliminary paper of 1933, he had described
in detail the wide array of physiological changes of metamorphosis in *Rhodnius prolixus*. In 1934, in a second lengthy paper,
he searched for the site of production of the hormone that he
presumed was responsible for such large-scale changes.[4] He

began by reminding his readers that *Rhodnius* invariably goes through five nymphal stages. In each of the first four moults it maintains the same body structures. Only during the fifth moult, when the insect becomes an adult, does it go through the striking transformation of its metamorphosis. As Kopeč had already demonstrated, beheading an insect larva did not necessarily kill it, but this drastic action had a profound effect on future development. To check this out for himself, Wigglesworth now beheaded examples of all five stages of *Rhodnius* larvae, using a ligature of thread in the larger forms and scissors in the smaller ones. The wounds caused by the beheading were sealed with molten paraffin wax.

In each stage, he confirmed Kopeč's 1922 observation that there was a critical watershed period in relation to the blood meal that determined the developmental effects of beheading. If he beheaded the nymph before this critical time, the moult did not take place; if he beheaded it after this time, it did. In the fifth Rhodnius moult, which normally took place about twenty-eight days after a blood meal – it seems likely that at this stage in his experimentation, Wigglesworth himself was the provider of the blood – he found he could pin down this critical time to between the sixth and eighth day. When he beheaded the insect larvae after the eighth day, the headless insects continued to digest their blood meals, excreted their waste, and went on to moult, following the normal timing and pattern. The insects beheaded before these days, though they survived for long periods, even as much as a year, failed to moult.

Wigglesworth was convinced that a hormone was involved in this process, and that it was released in or near to the head. It appeared the larvae decapitated early had not had the moulting signal conveyed throughout the body, while those decapitated close to the onset of moulting had already had the signal – the hormone – secreted and carried via the blood to the whole body. He was ready to design a new experiment that would further test this hypothesis. 'If this idea is correct' he said, 'the blood from an insect that has just passed the critical period should induce moulting in an insect decapitated before that period'.

The experiments became ever more bizarre. He beheaded two groups of nymphs, one (designated the primary) shortly after the critical period had passed and the other (the recipient) just twenty-four hours after feeding – and thus not expected to moult on their own. Then he used paraffin wax to join the two insect bodies together, neck to neck, so the blood could flow freely from one to the other – an experimental technique he called "parabiosis". Wigglesworth found that a number of the primary group moulted and, as they did, they induced moulting in their recipients. He went on to repeat the experiment many times with nymphs at different stages. In every case where the primary insect was decapitated soon after the critical watershed period, its blood induced moulting in a recipient decapitated twenty-four hours after feeding. He concluded that moulting was indeed initiated by a blood-borne factor, almost certainly a hormone.

But he also spotted something else, an observation that now seems precocious given the relative ignorance of developmental biology at that time. When he monitored the exact timing of metamorphosis in the beheaded and surgically-conjoined insects, the onset of moulting was accelerated in the recipient insect – it would take six days, for example, to reach a stage normally reached in nine or ten days – while moulting in the primary insect appeared to be delayed. These curious accelerations and delays lasted until the recipient insect had caught up with the development of the primary. It was as if the two insects were being integrated into one, with different parts of the conjoined body being brought into developmental alignment.

This led Wigglesworth to a brilliant deduction. 'It is inconceivable that a single external factor [i.e., hormone] should ensure this exact coordination', he wrote. 'These observations can only mean that the growing tissues themselves are communicating with and controlling one another by chemical means'. As if grasping the future importance of this thought, he added, 'This is a fundamental idea but it is beyond the scope of the present work'.[5] Indeed, this – another of Wigglesworth's famous asides – anticipated the discovery of signalling from cell to cell and from tissue to tissue, which would only come to be fully appreciated some half a century later by the new science of evolutionary development, or "evo-devo".

What he was now noticing might very well possess homologies with the groundbreaking discoveries being made in human hormonal research. Wigglesworth went on to suggest that the "moulting hormone" could only be secreted 'in the head or, more likely, by some organ innervated by nerves coming from the head'. But what possible organ could be the source? He focused on an organ known as the corpus allatum, which existed as paired structures, or corpora allata, on either side of the insect oesophagus, within the head. Named by the French biologist Charles Janet in 1899, the function of the corpus allatum was entirely unknown. Like the endocrine glands of vertebrates, it budded off from the epidermis at an early stage of development, and it was well supplied with nerves. Indeed there were many pointers to this being the insect equivalent of an endocrine gland, as it appeared to have close connections with a major nerve ganglion and through it with the brain. Wigglesworth began a series of careful dissection and staining studies, confirming that the corpus allatum was the only organ in the head to display a definite cycle of changes in the critical period after larval feeding. In doing so he also made an observation that seemed both intriguing and baffling. During the fourth nymphal moult, he detected a significant increase in the number of cells in the corpus allatum, with cell divisions, known as mitoses, being very numerous during the third, fourth, and fifth days after feeding. But during the final moult of the fifth nymph, the moult that gave rise to *Rhodnius's* metamorphosis, he saw no such increase in cells. On the contrary, the gland experienced no increase in size and no cellular mitoses were in evidence. To Wigglesworth, this suggested that the corpus allatum was active in normal moults and was then switched off, for some reason, during the final moult to the adult. The corpus allatum had to be the source of the moulting hormone.

Wigglesworth had also noted that moulting always followed a blood meal. Yet, it did not occur if the nymph was supplied with a steady trickle of blood instead of a single feast. This intimated that it wasn't the state of nutrition of the nymph but the massive abdominal distension of the blood feast that triggered the secretion of the moulting hormone. Further experiment confirmed this was the case. From this he deduced that the stretching of the abdomen sent a message to the brain, which in turn signalled the corpus allatum to get to work. This too he tested by cutting the nerve cord immediately behind the neck twenty-four hours after a full meal. Of thirty-two nymphs treated in this way, only six moulted.

Wigglesworth sat back and considered his findings. What then was the mechanism by which the fifth and final moult led not to a larger nymph but to the markedly different adult form? One possibility was that the hormonal control in the fifth moult was identical to that in all the other moults, but the body cells at this stage had developed some inherent difference, so they responded differently to the same hormonal stimulus. The other possibility was a change in the actual hormones themselves. He tested these alternatives with a new experiment, joining a fifth-stage nymph that had been decapitated after the critical period to a fourth-stage nymph decapitated twenty-four hours after feeding. Both insects moulted simultaneously, though the smaller, fourth-stage nymph suffered a premature metamorphosis into a smaller but otherwise normal adult. He repeated the procedure with first-stage nymphs, which would normally moult to a larger, second-stage nymph; instead they metamorphosed to diminutive adults. He could draw two possible conclusions: Either the moulting hormone of the final nymph was different from the hormone released in the earlier moults; or the hormone was always the same but the earlier nymphs also produced an inhibitory hormone which was absent in the final moult to the adult form.

Next he joined fourth-stage nymphs reared to an age beyond the watershed for hormone production to fifth-stage nymphs decapitated twenty-four hours after feeding. The fourth-stage nymphs, when they moulted, induced moulting in conjoined fifth-stage nymphs. But the fifth-stage nymphs did not moult to adults; they retained larval characteristics. Wigglesworth had his answer: 'The moulting hormone is the same at metamorphosis as at earlier moults. Therefore the absence of metamorphosis in the younger insects must be due to an inhibitory factor'.[6] This was an important breakthrough.

He had no idea where this inhibitory factor was produced, but one possibility was that it was secreted in the head. He even surmised that the corpus allatum might be the source of both the moulting and inhibitory hormones, through different types of cells stimulated as a result of different programmes controlled by the brain.[7]

Though he was spectacularly perceptive in his many observations, in this general conclusion of two hormones being secreted by the corpus allatum he was mistaken, at least in its universal application. A decade later, in his book *The Physiology of Insect Metamorphosis*, he would chide himself for this error, which had been made 'without good experimental evidence'. His conclusion was disproved by studies conducted in moths by Jean-Jacques Bounhiol and Ernst Plagge. Their experiments proved beyond doubt that the corpus allatum was the source of some important secretion – but it was not the source of the moulting hormone.

Still, Wigglesworth, ever the scientist, extended his parabiosis experiments to incorporate a wide variety of insects, including dragonflies, mayflies, locusts, and grasshoppers in addition to butterflies, moths, and, of course, the blood-sucking *Rhodnius*. Heads were joined together, resulting in double-bodied and double-headed conglomerates. Headless bodies were joined to intact heads of other bodies, resulting in insects with two bodies and one head. Finally, this frenzy of arcane experimentation gave him his most important clues to answering the mystery.

The first clue came when he noticed that if larvae of *Rhodnius*, in any nymphal stage from first to fourth, were decapitated soon *after* the critical period, they tended to undergo a precocious metamorphosis that led to adult characters in varying degree. The presence of an intact head was plainly important. 'There is clearly some influence exerted by the head that is preventing the realisation of imaginal [adult] characters latent within the cells of the larva'.

Wigglesworth demonstrated that if he joined the headless body of a first-stage *Rhodnius* to the head of a fifth-stage larva in the process of moulting, the tiny first-stage larva metamorphosed to a precocious adult. A hormone had to be involved, but it couldn't simply be the moulting hormone – otherwise the first-stage larva would have moulted not to an adult but to the second stage.

The penny dropped when Wigglesworth realised that what he was witnessing was not a stimulation effect but one of inhibition. If the corpora allata were removed from young silkworm larvae, they proceeded to spin cocoons and to pupate into miniature adults. Much the same thing was shown with the greater wax moth, *Galleria mellonella*, where removal of the corpora allata gives rise to tiny pupae and miniature adults. This did not happen if corpora allata from other larvae were trans-

planted into them. Conversely, if corpora allata were transplanted into full-grown larvae ready to metamorphose into
adult moths, metamorphosis was cancelled and larval development continued until eventually giant pupae and giant adults
were produced. On the other hand, applying a ligature behind
the head of honey bee larvae, at the right time, caused them
to moult directly to a form showing adult characters, including hair, skin, and the honey-gathering brush on the hind feet.
The corpus allatum certainly was producing a hormone, but
its action was not to induce metamorphosis but to prevent the
final change to the adult form.

Now his observations of a virtual "closing down" of the corpus allatum in anticipation of the final moult made more sense.
In choosing a fifth-stage larva that was already entering metamorphosis, he had caught the insect at a stage when the inhibiting hormone was turned off to allow for the final metamorphic
changes. A first-stage larva would normally have produced this
inhibitory hormone itself, and thus prevented precocious metamorphosis to an adult. But during decapitation, it had lost its
supply of the inhibiting hormone, so the same hormone crossing over into its circulation from the metamorphosing fifth
nymph had triggered the change to adult.

In a major paper published in January 1940, Wigglesworth
stated with certainty that the inhibitory hormone '… is secreted
by the corpus allatum in the first four nymphal stages'.[8] This
inhibitory effect was shown to cross species, genera, and even
families of insect. For example, the corpus allatum of the
tobacco moth, the domesticated silk moth, the mealworm beetle and the stick insect were variously shown to inhibit larval
to adult metamorphosis in the wax moth. In this same paper,
he declared he could no longer call his discovery the inhibitory hormone. 'In previous papers the "inhibitory hormone"
was so called because in its presence the production of imaginal
[adult] characters at moulting is suppressed. But now, in view

of its probable mode of action through the activation of the nymphal system at the expense of the [adult], it might be preferable to refer to this hormone as the "nymphal" or "juvenile" hormone'.[9] The world of biology accepted his argument and the hormone is now known by this latter name.

Wigglesworth could finally itemise the series of events that gave rise to insect metamorphosis. Stretching of the abdomen of the nymph set up a nerve reflex that travelled to the brain, very likely to the region of the brain identified as the pars intercerebralis, which only recently had been discovered by Bounhiol and Plagge.[10] Where he had earlier concluded that the brain next sent a signal to the corpus allatum, which secreted the moulting hormone, he now admitted he had been mistaken.[11] The moulting hormone came from some other, as yet unknown source, a mystery that had proved to be a considerable source of irritation to Wigglesworth, but which he was currently obliged to leave to others to resolve.

In 1931, the German entomologist V. Hachlow had shown that pupal development in the red admiral and the black-veined white butterflies was initiated by some centre located in the thorax.[12] In 1940, a Japanese biologist, Soichi Fukuda, identified the structure responsible: a diffuse organ made up of bead-like strings of cells wrapped around the trachea in the thoracic segment of caterpillars.[13] It was all the more ironic that this gland had been described as early as 1762, by the Dutch naturalist and engraver Pieter Lyonet. In fact, a number of naturalists had identified the same structure in different insects, in particular two fellow Japanese scientists, Kametaro Toyama, who had rediscovered it in 1902, and O. Ke, who, in 1930, had actually named it the "prothoracic gland".[14] In a series of elegant studies of the silkworm moth, Fukuda now demonstrated that if the larva was bound behind the prothoracic gland, the front end pupated normally and the hind part did not. But if he transplanted some prothoracic gland into the hind section after the ligation, pupation was restored. At some critical watershed period, the prothoracic gland released an "active principle" into the blood that controlled the moulting of the larva and the development of the pupa.

Perhaps we should pause to acknowledge how amazing it is that, at the very heart of the Second World War, scientists in Britain, France, Germany, and Japan, as well as several other countries, continued to engage in the experimental exploration of the great mystery of metamorphosis in insects. Indeed, on the night of Saturday May 10, 1941, at about 10.45 p.m., the London School of Hygiene and Tropical Medicine took a direct hit with a bomb. Nobody was killed, though there were some people working there at this late hour. A sixth of the building was destroyed, partly by explosion and partly by fire 'as it was not till morning that a fire brigade could be spared to come'. The basement of one full wing was severely damaged, as was the furniture in several departments. The devastation was so great its effects were still visible into the 1960s, but there was no interruption to Wigglesworth's experiments.[15] At the outbreak of war, anticipating such difficulties, he had taken the precaution of transferring much of his lab and equipment to his home in Beaconsfield, in the Chiltern Hills about twenty-five miles from the centre of London.[16] How difficult it was for the many other scientists working in such war-torn countries, one can scarcely imagine.

For Wigglesworth, dividing his work between London and Buckinghamshire, the scientific situation had also become a good deal more complicated than he had originally envisaged. Three separate organs had been identified as playing some role in metamorphosis: the pars intercerebralis of the brain, the corpus allatum, and the prothoracic gland. It now seemed inescapable that metamorphosis was not the result of a single active principle, but a succession of them. All that remained was to bring the three together into a cohesive, overall explanation.

It was at this point when, in the words of Bennet-Clark, 'there was a really awful blow to Wigglesworth's life. You've come across Carroll Williams?'

'Yes,' I replied.

'Carroll Williams represented, if I may say so, a major blow to Wigglesworth because he solved the puzzle of the moulting hormone.'[17]

The great experiment now extended to America.

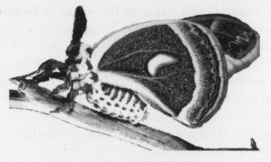

*Giant silkworm moth*

## 14

# Assembling the Jigsaw Puzzle

CARROLL MILTON WILLIAMS was born in Richmond, Virginia, on December 2, 1916. Like Wigglesworth, he became entranced with insects in his youth, so much so that during his undergraduate years at the University of Richmond he was appointed Curator in Insects to the university museum and helped the state entomologist of Virginia to identify thousands of local specimens brought in by the public. One species of butterfly took his fancy. This was the magnificent Diana fritillary, which is somewhat rare in America, being found only in a narrow geographic range extending from West Virginia to Missouri. Unlike other fritillaries, its large black and blue females are utterly different in colour to the smaller golden-orange and ultramarine males, so that only a skilled naturalist would even guess they are related. This led to Williams's first professional publication, a description of *Speyeria diana* and other butterflies native to Virginia, in 1937, when he was aged twenty, and just prior to his graduation.[1]

That autumn he enrolled as a graduate student at Harvard, where, with the help of fellow graduate student Leigh Chadwick and advice from the ingenious Harold "Doc" Edgerton of the Massachusetts Institute of Technology, he designed a stroboscopic apparatus that enabled him to measure the wing beat of the fruit fly. He registered between twelve and fourteen thousand beats a minute, a prodigious feat of muscular work ethic that could be sustained for as much as three hours – two million double beats. 'Finally', in the words of A.M. "Papp" Pappenheimer Jr, 'he succeeded in demonstrating, where others had failed, the neuromuscular network in the thorax that controls the wing beat'.[2] Williams completed what was described as a 'remarkable and brilliant' analysis for his PhD, granted in 1941, on "A Morphological and Physiological Analysis of the Flight of Drosophila".[3] Then World War II intervened. After joining the US Army, he was sent to Harvard Medical School 'to tool up on tropical medicine' and gained an MD, summa cum laude. 'Meanwhile', as he would later recount to a reporter from the *Boston Sunday Herald*, 'I had begun to work on a part-time basis on the physiology of insect metamorphosis'.[4] With the end of the war, and recently appointed Assistant Professor of Zoology at Harvard, he returned to his insect studies.

To contribute to the great experiment on insect metamorphosis, Williams needed a larger test animal than the tiny fruit fly. He decided on the giant silkworm moth, *Hyalophora cecropia*. The moth, which is North America's largest, is strikingly gorgeous, its body cloaked in cardinal red topped off with an ermine-white collar and its head crowned with huge feathery antennae. The velvety wings are a mosaic of patterned reds, browns, greys, and old gold, with flamboyant crescents and two arresting eye spots; the wingspan in the largest specimens extends to something close to fifteen centimetres. Like Fabre's great peacock, the adult moth does not feed: its sole purpose is reproduction. The female lays her neat rows of eggs on the leaves of sandbar willows and other host plants during early summer, and these hatch as tiny caterpillars of a waxy green colour. Feeding on trees and shrubs, such as the box elder and sugar maple – they will also devour the leaves and fruit of wild cherries, plums, and apples – the caterpillars must eat voraciously to grow to a prodigious ten centimetres long and three centimetres wide.

At this stage they are ready to spin their silken cocoons, which are attached along their length to a twig, hidden away in some dark and protected area. In many species of butterfly or moth, pupation and the final metamorphosis to the adult begins soon after the caterpillar enters the cocoon. But, critically for Williams's research, the *cecropia* moth follows a different strategy. When the pupa is sealed within its cocoon, it immediately enters a period of quiescence, called diapause, a process akin to hibernation in vertebrate animals. This tides the insect over the approaching winter.

The change from activity to diapause happens abruptly, as if an internal switch has been thrown. At the same time, all growth and differentiation comes to a standstill. But come the spring, when diapause ends, it's as if a second switch is thrown. Quiescence ends and a furious activity of destruction and reorganisation kickstarts within the pupa, leading to metamorphosis into the adult moth. These dramatic changes associated with diapause fascinated the thirty-one-year-old biologist.

When Williams began his research, nobody knew the nature of the two mysterious switches, or what threw them first to quiescence, then to dramatic metamorphic change. 'Whatever may be the inner mechanism for the induction and termination of diapause', he remarked, 'it must have the capacity to turn morphogenesis off and on in a most striking way'. This had so intrigued biologists that, in 1932, a French scientist named

Germaine Cousin cited 347 papers on the subject in a single review of the literature.[5] Now Carroll M. Williams was determined to solve it.

He began by establishing that a spell of cold weather was critical for the metamorphosis of his *cecropia* pupae. In nature this was provided by winter. But if he manufactured an artificial winter, say by chilling the pupae at temperatures of three to five degrees Celsius for a few weeks, adult development could be triggered by raising the temperature. Once he knew that temperature increase threw the second switch, it afforded him a means of investigating the switch itself. He invented a way of anaesthetising his insects with carbon dioxide, which was not only more humane but also permitted extensive and prolonged surgical manipulations without loss of blood or apparent damage to the pupae.[6] By 1946 he had performed 690 studies in 1200 pupae, leading to the first of a series of articles in which he searched for the factors controlling diapause.[7] In this opening paper he laid bare his thinking: 'If the termination of diapause is indeed accomplished by the action, within the previously chilled pupa, of a factor necessary for adult development, then it should be possible to demonstrate the organ in which this factor arises'.[8]

When he removed the brain from a pupa, he found that this led to a complete failure of metamorphosis. The larva persisted in suspended animation until death, which could take up to two years. But if he took the brain of a chilled pupa and placed it in the abdomen of a headless resting pupa at twenty-five degrees Celsius, the isolated abdomen metamorphosed to the adult equivalent. It didn't seem to matter that the recipient was headless any more than it mattered if the abdomen belonged to a different species of insect, such as the American saturniid silk moth, *Antheraea polyphemus*. In macabre fashion, the isolated abdomens would moult, attract a mate, and lay eggs. Non-chilled brains would not work. Thus he concluded, 'only one organ in the chilled pupa has the power to evoke development ... this organ is the brain itself'.[9]

The discovery that the insect brain controlled diapause fitted with Wigglesworth's findings of the importance of the brain during moulting in *Rhodnius*. It also fitted with the environmental cue Wigglesworth had established: while the brain of *Rhodnius* responded to the signal of larval abdominal distension, the *cecropia* moth brain responded to an increase in ambient temperature. These results further fitted with the findings of other biologists who had shown that the brain was equally important during the pupation of completely metamorphosing butterflies and moths. Williams concluded his paper with the teaser: 'What then is the nature of the factor arising in the brain that so spectacularly evokes the changes of metamorphosis in the chilled pupae?'

One possibility was that the implanted brain worked by some direct effect on the host tissues, converting them from dormancy to activity. But there was another, more intriguing option: the transplanted brain – perhaps the pars intercerebralis identified by Jean-Jacques Bounhiol and Ernst Plagge – might act indirectly, by producing a hormone that stimulated a more peripheral organ in the recipient larva to produce the moulting hormone.

To evaluate this, Williams needed to chop up individual pupae into smaller fractions. But he knew from experience that division of the stout silk moths by ligatures, such as had been pioneered by Kopeč, had not met with success. Even less appealing was the technique, first attempted in other insects, of slicing the unfortunate pupae into traumatised parts. Such brutal procedures were condemned to failure because chopping through the mid-gut filled the blood with lumps of tissue, and this embolised the heart with solid debris, killing the test animals.

Williams set about inventing a less brutal method, carried out under carbon dioxide anaesthesia, in which he carefully dissected out and separated the lower six abdominal segments of brainless diapausing pupae from the front end of the animal, while keeping the mid-gut intact. The gut could be brought forward into the front end or dissected out and removed entirely. This led to a controlled severance of the front from the rear half of the pupae. He then sealed the cut ends with paraffin wax, in which he embedded a microscopic cover slip. The cover slip had a central aperture through which he could inject Ringer's solution, a mix of salts dissolved in water, so as to maintain physiological homeostasis and keep out air. The hole, which was sealed with more paraffin wax, could be reopened at any future stage to allow him to operate through it. In the meantime, the cover slip offered him a window for inspection.

Despite Williams's use of anaesthesia and his skilful dissection, two-thirds of the isolated pupal parts died within days. But the remainder survived for lengthy periods. Into each of the separated halves of the pupae he now implanted a chilled *cecropia* brain. The front ends metamorphosed normally to the corresponding front end of a lively moth; the isolated abdomens, on the other hand, remained undeveloped. He found he could transplant as many as six chilled brains into a single abdomen without inducing a jot of change, even though these abdomens survived for as long as eight months. In his words, 'This difference in response might be explained if the [front] fragment possessed a second developmental centre that was lacking in the posterior fragment'.[10]

More experiments followed. Isolated abdomens were grafted to brainless pupae, or to the front end of the severed bodies of brainless pupae. Graft and host joined up to grow together, but they failed to metamorphose. When he introduced a chilled brain to either combination, however, the entire parabiotic construction metamorphosed to the corresponding parts of the adult.

He moved on to grafting individual internal organs. 'This search was vastly aided by the publication of Fukuda's paper in 1941', as Williams would subsequently acknowledge, referring to the Japanese biologist who had discovered that the prothoracic gland played a fundamental role in metamorphosis of the commercial silkworm, *Bombyx mori*.[11] Williams next confirmed 'the great significance of the prothoracic glands' in *cecropia* when he found that isolated abdomens metamorphosed readily when provided with chilled brains plus two pairs of prothoracic glands.[12] This was a key realisation: the prothoracic glands were essential for metamorphosis, providing the ingredient that Vincent Wigglesworth had acknowledged to be missing in his overall assessment. Moreover, Williams could now take understanding a vital step further: 'From these observations it may be concluded that the brain exerts a controlling action on the prothoracic glands'.[13]

In further studies, published in a series of papers extending into the 1950s, Williams showed that metamorphosis in *cecropia* involved three separate steps: 1) the secretion of a hormone by the prothoracic gland, which would fit with Wigglesworth's proposal of a moulting hormone; 2) the production of this hormone under the control of a master hormone, then unknown, secreted by the brain; and 3) the *absence* of a third hormone secreted by the corpus allatum – Wigglesworth's juvenile hormone.[14]

Williams went on to become Professor of Zoology at Harvard, by age thirty-six having established a distinguished reputation in entomology. In 1955, two years after his professorial appointment, he travelled to Cambridge, on a two-year Guggenheim Fellowship, where he worked in Wigglesworth's university lab. The two men did not become close associates, in the memory of others who worked there. Not that there was open hostility between them. Both were too gentlemanly for any unpleasantness, but perhaps professional rivalry and the differences in personality may have proved too great an obstacle.

Wigglesworth, by all accounts, appeared shy and withdrawn, never permitting a free and easy friendship with any colleague. His retiring nature even extended to his children, who would joke that they should have been insects, since only then would he have paid them sufficient attention. He was not devoid of humour, however: in his famous weekly teas, his cackle of a laugh was unmistakable, and a dry humour is noticeable in his more philosophically or publicly directed books and papers.

Williams, in the words of "Papp" Pappenheimer, was 'a traditional Southerner, a man of charm and unfailing courtesy, who kept his rich Tidewater accent untainted by clipped Yankee tones throughout his life'. Simon Maddrell, who was a youthful postgraduate researcher in Wigglesworth's lab at this time, would recall him as a genial extrovert, who walked the Cambridge streets in a "ten-gallon" cowboy hat. Where Wigglesworth worked for the most part in solitary intensity, nearly all of Williams's research papers were collaborations with other workers, often his postgraduates. He appears to have been more extravagantly sociable, with 'humourous overtones' even in his scientific papers, a characteristic that emerged as early as his PhD thesis. Indeed, one of his graduate students recalled, 'When I think of Carroll's achievements, I am overwhelmed by memories of hilarious events and merry times … life in his lab was usually such fun and we all shared so many laughs'.[15] Not everybody came away with such a genial impression, though. According to his subsequent research associate, Lynn Riddiford, Williams had a very strong personality and had forceful opinions on the relevance of his work and that of others. 'There were people who liked him and people who didn't', she said.[16]

In fact, when one looks a little deeper than the differences in personalities, there were striking commonalities between the two scientists. In both we discover an unusually intense and pure dedication to their science, evident in both their careful planning of experiments and in their massive outpouring of work. Only later, when he himself had succeeded to Wigglesworth's chair at Cambridge, did Maddrell come to see that, for Wigglesworth, 'science was to a large extent his life. He devoted himself to it in a very natural way, not out of duty but driven by a great curiosity about natural things and how they worked, and driven by a great desire to explore this whole new area of insect physiology'.[17] An American colleague might equally have been talking of Williams. This is exactly the determination one might expect to find in a winning Olympic athlete. Where others met with more limited success, the single-minded determination of Wigglesworth and Williams helped them to succeed at a deeper, altogether more profound level.

I was struck, in considering this great experiment, by the fact that Wigglesworth, who had pioneered the discovery of juvenile hormone, had never taken his researches to their logical conclusion. Rather he appeared to have abandoned this to Williams. In questioning Maddrell about this, he acknowledged, 'If he had continued his deep involvement with metamorphosis, one would have expected him to move on from Rhodnius to the endopterygotes' – that is, fully metamorphosing insects.

I pressed him a little. 'He obviously never felt it necessary to do that?'

'No. He took it to the death of what interested him and then he went on to something else.'

'Perhaps he just felt that he had taken it far enough when it got down to the more tedious biochemical extrapolations?'

Maddrell shook his head. 'I don't think that was the motive.

I think I understand what it was, at least partially, now. His attitude would have been: "There's just some other mystery I'd like to get on to".'

'I can see that it was a fantastic position to be in – when there was so much that was unknown.'

'Absolutely. He created the field.'

*Culex mosquito larva*

## 15

# Ecology's Magic Bullet

IN 1954, A GERMAN BIOCHEMIST, Peter Karlson, working at the University of Tübingen, together with Adolf F.J. Butenandt of the Max Planck Institute for Biochemistry in Munich, crushed vast numbers of pupae of the silkworm moth, *Bombyx mori*, to accumulate almost five hundred kilograms of a crude homogenate. From this they isolated less than thirty grams of a crystalline extract of purified moulting hormone.[1] On analysis, they found that, like the substances secreted by the adrenal glands in humans, it was a steroid hormone. They named it ecdysone.

At about this time, Carroll Williams decided to switch his research to finding a method for extracting and purifying juvenile hormone – a fitting turn, since the inhibitory hormone had first been detected by Wigglesworth, and it was in Wigglesworth's lab that Williams made his breakthrough. Williams began by confirming that excision of the corpora allata from chilled *cecropia* pupae had no effect on development into normal fertile adult moths. One would expect inactivity of the corpus allatum at this pupal stage, but when he tested the corpus allatum in the adult moth, he found, to his surprise, that it was more active than at any stage in the moth's life history. Indeed, the highest concentrations of juvenile hormone were to be found in the abdomens of adult male silkworm moths.

During his stay in England, Williams made his first crude extracts, using petroleum ether, by pulping large numbers of male abdomens. On his return to Harvard, he continued to gather increasingly pure isolations, work that carried over into the following decade. In 1964, he was joined by Karel Slama, who had been conducting an unrelated line of research in his Prague laboratory rearing the firebug, *Pyrrhocoris apterus*, from fertilised eggs. This common and widely distributed insect takes its name from its flame-red against charcoal colouration, which resembles glowing coals, a resemblance that can be very striking when it forms dense clusters of tens or hundreds of individuals. However, when Slama attempted to replicate this work at Harvard, the fertile eggs failed to mature, simply forming supernumerary larvae. The difficulty was eventually traced to the paper he had packed into the rearing jars. The startled entomologists then realised the plant kingdom had beaten science to some handy applications for insect metamorphosis.

When one considers how vulnerable plants are to attack by the multitude of animals that prey on them, one marvels at their endurance. Plants have, of course, evolved physical deterrents of their own, such as thorns, stings, and noxious chemicals; contrary to some popular misconception, nature is far from benign. Indeed, as Dame Agatha Christie was well aware, some of the most lethal poisons in the world derive from common plants in the wild, such as deadly nightshade. Some of these poisons help plants to defend themselves against insect predation by interfering with insect metabolism. Williams and Slama discovered that natural production of an analogue of juvenile hormone by many different plants was an important ingredient in this protective repertoire. In time, juvenile hormone analogues would be isolated from a variety of common plants such as the balsam fir, the pepper tree, sweet basil, and sedge. Topical applications of these analogues to final-stage insect larvae were found to block metamorphosis to the adult, resulting in a supernumerary moult to a giant immature larva, much as the entomologists had discovered in their experiments. These juvenile hormone analogues also found their way from plants into manufactured paper, including the pages of the major broadsheets *and Science* and *Scientific American* magazines, which were shown to inhibit the metamorphosis of those same firebug larvae. When Slama replaced the paper in the rearing jars with the highly purified Whatman filter paper, metamorphosis proceeded normally.[2]

Williams was not slow to realise the practical potential. Juvenile hormone, or a closely derived analogue, could be useful as an insecticide. Affected larvae would fail to mature, cutting off the reproduction of unwanted pests.

In 1956, he wrote a letter to *Nature* in which he described the discovery of juvenile hormone. In it, he coined the catch phrase the "paper factor", noting that '... in addition to the theoretical interest of the juvenile hormone, it seems likely that [the hormone] will prove to be an effective insecticide. This prospect is worthy of attention because insects can scarcely evolve a resistance to their own hormone'.[3]

At about this time, Williams had begun working with Lynn Riddiford, who, in an interesting historical footnote, was described in the 1960s as 'one of the few women scientists in Harvard history to reach professorial rank'.[4] The two scientists were having difficulties in mating the saturniid silk moth, *Antheraea polyphemus*, in the laboratory. Realising that the larvae of *Antheraea* feed on the leaves of oak trees, they added a few oak leaves to the cages, whereupon the female appeared to secrete her pheromone and mating preceded normally.[5]

In a Boston *Sunday Herald* article, in August 1959, Williams was quoted as hoping the chemical formula for juvenile hormone would be written by the end of the year, which should allow not only its synthesis but also the manufacture of a series of chemical analogues and derivatives. These notions provoked some humourous interludes with the lay public, who equated juvenile hormone with the "elixir of youth", assuming it might act, in Peter Pan fashion, to retard ageing in humans. The newspaper quipped: 'Crowds of fascinated people cannot keep from guessing what the end results of the Harvard biologist's youth hormone experiments may be'. The bemused Williams received torrents of telephone calls, cables, and letters from individuals 'seeking to splash in the fountain of youth'. He also recounted stories of cosmetic manufacturers who '… plead with me to have a little of the stuff in crude form to put into beauty creams and lotions.'

'If you could drink the hormone yourself,' a reporter asked him, 'when would you take it?'

'Right now!' was the tongue-in-cheek retort.

In a series of papers extending into the next decade, Williams and his colleagues went on to demonstrate the production of juvenile hormone by the corpus allatum and its role in the endocrine control of moulting, pupation, and adult development in the *cecropia* silkworm. In 1965 came the crowning achievement of its extraction and purification.[6] As both Wigglesworth and Williams had foreseen, this had consequences that extended far beyond entomology, including practical applications that are increasingly relevant to medicine and agriculture today.

Thanks to these pioneers in this great experiment, we now know that the moulting process begins in the neurosecretory cells in the insect brain, which send messages along nerves to a pair of tiny organs, the neurohaemal organs which are located very close to the corpora allata. These in turn release a store of prothoracicotropic hormone (PTTH), which is carried by the blood stream to the prothoracic gland, where it stimulates the production of the moulting hormone, ecdysone. The close anatomical proximity of the neurohaemal organs to the corpora allata explains the initial confusion in Wigglesworth's results, and his thinking. To add to the confusion, in butterflies and moths PTTH is actually stored in the corpus allatum, so its removal would, in the Lepidoptera, cut off the stimulus to production of ecdysone by the prothoracic glands.[7]

Each moult of the insect larva is triggered by one or more pulses of ecdysone. A first pulse of neuro-electrical excitation from the neurosecretory brain cells induces a small rise in ecdysone in the larval blood. This elicits a change in cellular commitment – in other words, it acts as a kind of developmental primer. A second, larger pulse of ecdysone sets the ball rolling for the full changes in tissue and cell differentiation associated with the moult. The location of the priming signal in the brain allows environmental cues to influence the timing of moulting, for instance, the abdominal bloating in *Rhodnius* after a blood meal as discovered by Wigglesworth. Another example of control through the same basic mechanisms is seen in the silkworm moth, *cecropia*, the subject of Carroll Williams's groundbreaking experiments, where PTTH secretion abruptly ceases after the pupa is formed. The pupa can thus remain in the suspended state of diapause throughout the winter, after which the temperature rise cues its final moult to the adult. The in-built mechanisms of diapause also explain how *Rhodnius larvae* lived for up to a year after decapitation. An understanding of such processes is important in the silk industry, where the silk-spinning pupa is artificially primed by two weeks' exposure to cold.

We also now know that Wigglesworth was right in his earliest experiments on the inhibitory effects of the corpus allatum:
this does indeed produce juvenile hormone, which blocks metamorphosis in the younger nymphs, larvae, grubs, maggots,
and caterpillars of all metamorphosing insects. Wigglesworth
demonstrated this in flamboyant manner when he painted his
initials in juvenile hormone onto one of the abdominal segments of a fifth-stage *Rhodnius* nymph before metamorphosis.
When metamorphosis was complete, the initials "VBW" were
permanently etched in larval colours and texture on the differently coloured and textured abdominal skin of the adult – an
experiment, based on a single test subject, that won Henry Bennet-Clark's admiration.

Posterity has also confirmed that when the larva is fully
grown, the corpus allatum switches off the production of juvenile hormone. This removal of the inhibitory agent results in
the metamorphosis to the adult in incompletely metamorphosing insects and in the metamorphoses from larva to pupa, and
from pupa to adult, in completely metamorphosing insects.
Under the stimulus of the moulting hormone, ecdysone, developmental genetic programmes intrinsic to this stage of the larval and pupal tissues effect the changes.

This great experiment into insect metamorphosis must rank as
one of the most exciting and instructive in twentieth-century
biology, a majestic demonstration of the power of deductive
reasoning and scientific experiment in unravelling a fundamental mystery of nature. For medical epidemiologists working against lethal epidemics such as malaria, trypanosomiasis,
yellow fever, and dengue, all of which are spread by insects,
such understanding offers the potential for biological control
of the disease.

In 1966, just a year after he and his colleagues had first reported the chemical extraction of juvenile hormone, Williams demonstrated the fatal effects of a synthetic juvenile hormone on the larvae of the yellow fever mosquito, *Aedes aegypti*.[8] A year later he wrote a review article for *Scientific American* on the practical applications of juvenile hormone analogues for agriculture as well as human health.[9] That same year he and Riddiford showed the effects of juvenile hormone analogues on the embryonic development of silkworms.[10] In an interview with the *Boston Globe*, the two scientists explained the future ecological potential of such extrapolations of their metamorphosis experiments: 'It seems possible to develop hormone insect-killers that will affect only one specific kind of insect pest. Laboratories in several countries are working on this. And the insects probably will not be able to develop the kind of resistance that they build up against insecticides such as DDT'.[11] In this, Williams appeared to be predicting the "ecological magic bullet", a perfect treatment, much as the nineteenth-century German immunologist Paul Ehrlich had predicted chemical cures for infections before the era of antibacterial and antiviral drugs.

Today, hundreds of artificially constructed juvenile hormone analogues have been produced. These and a variety of other extrapolations have made possible environmentally sensitive pest management through hormonal disruption of insect development, metamorphosis, reproduction, diapause, and other aspects of insect behaviour.

One such juvenile-hormone derivative is methoprene, a general-use pesticide that is sold under a variety of trade names. Sufficiently environmentally sensitive to be classed by the US Environmental Protection Agency as practically non-toxic, a mere five grams of methoprene will cover an acre of mosquito floodwater ground. It seems likely that the wide diversity of insect metamorphic and reproductive strategies will, in time, offer an ever increasing range of highly targeted solutions to infestations.

I discovered a remarkable example of this developing biotechnology when I visited Chris Curtis and his colleagues at the London School of Hygiene and Tropical Medicine. Curtis is an expert in malarial research and control. His research interest is the malaria-carrying *Anopheles* mosquitoes, which he breeds in the underground insectaries, extending out under Gower Street. He has offered himself as live bait while conducting mosquito counts in African villages, sitting out of doors all night with his trouser legs rolled up – a risky dedication to duty that, on one occasion, gave him a bout of falciparum malaria. You could say he knows his subject intimately. His office wall is adorned with a map of Africa, with coloured pins marking the sites of different strains of *Anopheles* eggs that have been ferried to London for breeding experiments.

Malaria has long been a scourge in Sri Lanka, as in many other tropical and sub-tropical areas of the world. But in 1967, thanks to widespread spraying with DDT, the medical authorities there almost eradicated the disease, with just seventeen cases reported that year for the entire country. All seventeen cases were found in the central area of Elahera, where there is a tradition of open-cast gem mining. Under licence from the State Gem Corporation, the miners dig shallow pits by hand in search of gemstones such as emeralds. They are required, under the regulations, to fill in the pits after they have been exhausted, but in practice many pits are abandoned to fill up with water during the rainy season. These become breeding grounds for *Anopheles* mosquitoes. In the 1990s medical investigators estimated that a nomadic population of twenty-five thousand to fifty thousand people were still engaged in gem mining, and the numbers of shallow pits that could form potential breeding habitats for mosquitoes were estimated to be between 247 and 370 per hectare. To make matters worse, the huts the miners occupied were made out of woven palm leaves and polythene, which gave no physical protection from mosquito entry and proved unsuitable surfaces for insecticide spraying.

Traditional preventive measures no longer worked. The result was a resurgence of malaria centred on the gem-mining areas and amounting to hundreds of thousands of cases annually, and a great many deaths. Over the years 1993 to 2000, one of Curtis's postgraduate students, Amara Yapabandara, began working with experts from the Sri Lankan anti-malarial campaign. With Curtis's help, they set up a trial of the juvenile hormone analogue, pyriproxyfen, in an attempt to eradicate the *Anopheles* population in the mining area.[12]

As her test area, Yapabandara chose a cluster of eight villages, mapping the area to show the geographical distribution of the houses and any potential breeding places. The location of every house, path, and major water body was surveyed by counting paces along compass bearings. Each house was given a geographic reconnaissance number, which was marked on the door. Within a zone of almost one mile around each village, all the gem pits were also numbered, and mosquito populations in the study areas were estimated based on a variety of sampling methods. Information on malaria cases was gathered from four different field clinics and from the outpatient departments of the local government hospital and dispensary. This involved a huge community operation, with village education and discussion, and periodic mass blood screening in addition to the monitoring of infected patients. Of the eight chosen villages, four were assigned to be treated and four assigned to be controls. In the four chosen for treatment, Yapabandara sprayed every pit, amounting to several hundred pits per village, together with any other potential sources she had spotted in the initial survey, with pyriproxyfen twice a year. Mosquitoes were systematically counted and cases of malaria tabulated, and the results of the treated villages were compared to the controls.

'It worked beautifully,' Curtis told me. 'There was a dra-matic reduction in mosquito counts in the treated villages, and such an effective reduction in malaria that ... it clearly showed that pyriproxyfen was sufficient to suppress malaria in the treated villages compared with the controls.'

As this practical application implies, this line of research continues in the field and in many laboratories throughout the world. According to a lucid summary by H. Frederik Nijhout of the Department of Biology at Duke University in North Caro-lina, 'virtually every aspect of the post-embryonic development of insects is controlled by hormones'.[13] This includes metamor-phosis *per se*; the timing of developmental processes; the variety of moulting cycles; the fates of individuals among the social insects, determining castes of ants, bees, and termites; and the distinctive seasonal forms seen in many species. The hormones involved in this huge variety are eminently simple: a handful of ecdysones and juvenile hormone. These are responsible, as Nijhout explains, 'for virtually the entire panoply of events'. Such is the legacy of the great experiment undertaken by Wigglesworth, Williams and their colleagues across many countries, all of whom contributed to the investigation of metamorphosis by experiment.

But understanding the physiology of metamorphosis in insects, elegant as this hormonal control has proved to be, is not synonymous with understanding the evolutionary origins of metamorphosis, whether in insects or in its wider extrapolations to development across the entire animal kingdom. To do this we must cast our net wider. If there are commonalities in the evolution of metamorphosis throughout the animal kingdom, we should bear in mind what we have learned from the study of the physiology in insects, while looking once again to the oceans, where we discover a great many phyla embracing myriad varie-ties of metamorphosis more baffling and varied than even the strange and beautiful world of the insects we have beheld. It is to the marine world, the real cradle of life, that we must return.

# Part III
## New Perspectives

*Yet to admire only our own successes, as if they had no past (and were sure of the future), would make a caricature of knowledge. For human achievement, and science in particular, is not a museum of finished constructions.*

Jacob Bronowski,
'The Ascent of Man'

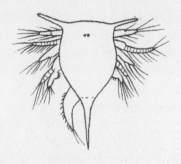

*Nauplius larva of barnacle*

# 16

# On the Steps of York Minster

AT NINE O' CLOCK on the evening of January 25, 1991, in the freezing cold of the Yorkshire winter, three figures waited on the steps of York Minster for the arrival of an American editor named Greg Payne. Don Williamson, wearing padded Rowans and huddled under a car rug in a wheelchair, was shivering. He was accompanied by his wife, Enid, and Mary, his sister, both also shivering despite neck scarves and winter coats. They had arrived early and Payne was late. Mercifully it had not snowed. 'We didn't know Greg by sight. We had never met him. People were coming up to speak to us, assuming we were keeping some special vigil or something, though nobody offered us any money.' The sense of humour was quintessentially Williamson. And now, with the prospect of his visitor, the marine biologist had rediscovered his sense of hope and purpose.

While still hospitalized in Douglas, Williamson had received another rejection of his book. It had already been rejected by six publishers when Lynn Margulis intervened to encourage two editors she knew to take it seriously. The first of these was an editor-at-large at Yale University Press to whom Williamson sent a copy of the incomplete manuscript. The editor had responded enthusiastically but added that he would need the approval of his editorial board when it next met. Sadly, in the interim, the editor became seriously ill and his successor promptly declined the project.

As this was going on, Williamson was facing other difficulties that for many would have proved insurmountable. After his stroke, he had completely lost the ability to speak, a condition known as aphasia, which was accompanied by the inability to read. Over time he had made a partial recovery. His aphasia had also been accompanied by dysgraphia, an impaired ability to write. When the Yale University Press editor's rejection arrived, Williamson initially had difficulty comprehending the letter. When he finally managed to decipher it, he was devastated by its content.

At this time he shared a two-bedded ward with Robert, a seventeen-year-old boy admitted with recurrent arthritis who was studying for his A-level examinations. The sixty-eight-year-old scientist and the youth soon struck up a friendship. Robert became fascinated by Williamson's theory of metamorphosis, though it was an enormous struggle for Williamson to explain it to him, between his difficulties with speech and Robert's limited knowledge of biology. Nevertheless, little by little, over the fortnight or so during which they shared the room, Williamson outlined something of biology, the prevailing evolutionary principles in general, and the challenging nature of his own theory in particular. The exercise proved therapeutic. 'Robert asked me intelligent and searching questions and this made me work through the whole thing and even enlarge on it. It was very hard work but I think it was worth it because it forced me to relearn how to put my thoughts into words.'

As Williamson's speech improved, his ability to read and write also began to recover. But he was left with a paralysis affecting his right hand, so he set about teaching himself to type using only his left one. For the previously right-handed scientist, this was made all the more difficult by the fact that he had lost the spatial memory of the typewriter keyboard. He had to re-educate mind and hand, like a child at infant school first learning the alphabet, but he persisted, day after day, and finally recovered this ability, too.

All hope for publication of his book now focused on the second of the two editors Margulis had recommended. This was Gregory W. Payne of Chapman and Hall in New York, for whom Williamson was waiting on the cathedral steps in York, and to whom Williamson had written on his return home to Port Erin. 'In retrospect it seemed rather ironic, because I had already submitted the book to Chapman and Hall in London, who had rejected it, saying, "When your theory is better known, we will reconsider it".' Of course, from Williamson's perspective, the whole purpose of publishing the book was to get his theory better known.

Once back home, and still affected by his stroke – his recovery would in the long term prove to be a partial one – Williamson switched from a typewriter to an early model Amstrad home computer and began, all over again, describing his theory and publication hopes to Payne. To Williamson's joy and relief, Payne wrote back: 'Certainly, we will publish it.'

As it happened, Payne was planning to visit Yorkshire, where Williamson and his wife had gone to visit with his sister. It was Payne who suggested the steps of York Minster as a meeting place. So it was that Williamson, his wife, and his sister were waiting there on that bitterly cold January evening. Payne arrived half an hour late, accompanied by a friend. All

five retired to the nearest hotel, where in more congenial surroundings the editor duly produced a publisher's agreement,
which Williamson signed on the spot. Payne comforted Williamson prior to leaving, 'We'll try and have it out for your
seventieth birthday.'

In the event, Don Williamson's book, *Larvae and Evolution:
Towards a New Zoology*, was published by Chapman and Hall in
1992, six months after he turned seventy.[1] He dedicated it to
Enid, who had not only supported and sustained him throughout his difficulties but had also drawn many of the illustrations
– and there were a great many of these, both wondrous and
arcane. Anybody reading through the book for the first time will
be struck by its clarity of thought and exposition; as a book of
science, it is, in the very best sense, quite beautiful. Under normal circumstances, it would be considered a laudable achievement. In the circumstances of its genesis and publication, it was
nothing short of remarkable. Lynn Margulis and Fred Tauber
wrote the foreword, in which they described Williamson's difficulties in getting his ideas published and made apparent their
opinion that Williamson's theory deserved the spotlight of scientific scrutiny and the courtesy of a proper hearing.

Williamson's talk in Boston in 1990 had provoked a heated
reaction among some of the professional biologists who made
up his audience. A book is a more comprehensive vehicle than
a lecture, allowing more detailed argument. In his preface,
Williamson attempted to mollify the inevitable criticisms by
explaining that he was perpetrating only a minor heresy. By
and large he agreed with, and admired, Darwin. He accepted
Darwin's idea of evolution as the explanation for the diversity
of life and he believed natural selection had played a fundamental part in that process. What he disagreed with, in essence,

was the universal insistence on a classical branching-tree depiction of speciation.

Darwinians think the formation of new species comes about through branching from an existing twig on the tree of life – a linear process of divergence. But Williamson contended that the only logical explanation for what he had observed in the evolution of some animals and their larvae was that occasionally some branches had actually met and fused at some point during their evolutionary history. This reunion of branches, as opposed to the linear process of branching, constituted a "reticulate" pattern of evolution. Moreover, in his opinion, reticulation was not confined to neighbouring branches, such as those species within a genus, but extended much wider, to include the fusion of very different evolutionary groups. This resulted in the coming together of whole pre-evolved genomes – the summing of all their genes – into new life forms. To Williamson's way of thinking, his theory did not refute Darwin; instead, it might best be interpreted as a modification of Darwinism as most biologists currently interpreted it.

In his book, he made his case by exploring the links between the development of various animals and their evolutionary history, providing many examples where hybridisation might have resulted from reticulate fusions of different evolutionary pathways to give rise to some of the present-day manifestations of metamorphosis. He began with the crustaceans, explaining the anomalies he had observed, before moving on do the same for the entire phylum of the echinoderms. The deeper he looked at this colourful and varied group, including sea urchins and starfish, the stranger the life histories, and by implication the evolutionary origins, of these familiar creatures appeared to be.

We have already seen that all living adult echinoderms are radially symmetrical in the horizontal plane, a quality they share with flowers. In the vertical plane, they all have an "oral" underside, where you find the mouth, and an "aboral" top side, where, in most echinoderms, you find the anus – but there is no back or front, no left or right. The outer surface is armed with the calcareous spines typical of their phylum. The underside also bears soft flexible tube feet, which enable the animals to creep over the ocean floor. The tube feet are connected to an internal system of fluid-filled canals, the water vascular system, which can extend and contract the feet. In many species the tube feet are also fitted with suckers to enable them to attach to the ocean floor and to capture prey.

The echinoderms have a tube-shaped gut that runs through the internal cavity known as the coelom, into which the gonads protrude. The coelomic fluid is circulated internally by cilia. With only rare exceptions, eggs and sperm are shed through this same ciliary action into the sea, where they are fertilised and left to survive on their own. A single starfish can extrude as many as two million eggs in one spawning. Startling as this might seem, it is no more prodigious than the numbers of sperm ejected in a single male emission in humans.

A striking and little realised similarity is also found in women, where the ovaries protrude into the human version of the coelomic cavity and the eggs are shed into it, whence they make their way into the collecting tentacles of the Fallopian tubes, then pass along the tubes and down into the uterus, which has an open connection, through the vagina, to the outside environment. In this way, we can envisage our distant oceanic past, for when women bathe in the ocean, their coelomic cavity is connected to the ocean, however tenuously, and in such circumstances human eggs, if not fertilised and implanted in the uterus, would also be shed into the sea.

Echinoderms possess a simple heart, which circulates their equivalent of blood through a system of haemal channels. But other aspects of the echinoderm's inner structure are altogether strange when compared to our bilaterally symmetrical inheritance. Echinoderms have nothing resembling a brain. Instead, nerves radiate from a ring-like structure around the mouth and branch off to all parts of the animal in contact with the outer environment. All the same, this is enough for these intriguing creatures to control and coordinate the necessities of daily life, sensing prey, making purposive movements towards it, consuming their food, and excreting their waste.

There are considerable variations on this basic theme among the different classes of echinoderms. Where starfish creep on tube feet equipped with suckers, brittle stars use long, snake-like arms for locomotion. Curiouser still, where starfish have a gut serviced by a mouth at one end and an anus at the other, brittle stars have a single opening to and from their digestive tract through which they both ingest their food and void the waste contents of their gut, as we saw with Kirk's brittle star.

Sea urchins, heart urchins, and sand dollars have no radially distributed arms. Instead, the body is enclosed within a rigid shell of closely fitting plates that support movable spines and openings through which the tube feet protrude. Like their star-fish cousins, sea urchins have both a mouth and an anus. Their coelom-placed gonads are considered a delicacy in some countries, where they are farmed. Sea urchins and sand dollars are equipped with five teeth, which enable them to graze food from hard surfaces. These teeth are supported by a skeletal structure known as Aristotle's lantern, as the philosopher dissected and first described them. In contrast, heart urchins are toothless and so are not capable of such grazing. They live in tubes in the sand, from which they move particles of food to their mouths using modified tube feet.

Sea cucumbers are different again, with member species shaped somewhat like the vegetable that gives them their name. Sea cucumbers are also consumed as food in certain parts of the world – though they taste rather differently from the vegetable. Feather-stars and sea lilies resemble a beautiful mass of colourful feathers or clusters of long-stemmed flowers. Though both feather-stars and sea lilies are animals, they have phases in their life cycle when they are tethered to the ocean bottom or to driftwood and are among the most gorgeously decorative residents of coral reefs.

Finally there is a strange, newly discovered class of echinoderms, the sea daisies, which resemble their namesake's flowers, with tiny heads a centimetre or so in diameter. These are known from three species within the genus *Xyloplax*, which were discovered in sunken, waterlogged wood in deep water off the coast of New Zealand.

All of these diverse classes of echinoderms, with their differing life cycles, were considered by Williamson in relation to his larval transfer theory and the cohesive evolutionary synthesis he devised in his book.

Darwin had adopted a linear theory of evolutionary progression, with natural selection operating, small steps at a time, over vast periods. This incremental and linear process operated at every stage of the individual organism, from fertilised egg to embryo, and then right through to adult. Today, most evolutionary biologists still espouse this view, which allows for evolutionary innovation to take place at any stage of the life cycle.

But Williamson did not subscribe to any all-embracing interpretation of Darwinian linearity. 'The theories I am now putting forward imply that this assumption is frequently invalid', he noted in his book. He proposed that the larval body plan and the adult body plan sometimes derived from entirely different lineages. If true, this had major implications for the evolutionary history and genetic inheritance of marine life forms. Evolutionary biologists from Darwin onward had looked at the structures of embryos and larvae as important pointers to the evolutionary history, and place on the tree of life, of the adults. Darwin himself had pioneered this approach in his famous examination of barnacles.

As we have seen, the molluscs include a wide diversity of forms and lifestyles, from the familiar shellfish of beaches and seashores, such as mussels and whelks, to the seagoing octopuses and jet-propelled nautiluses. Most molluscs have a soft body, surmounting a muscular foot, and are protected by the secretion of an external shell. On superficial examination, barnacles also appear to fit this bill. But when Darwin studied their metamorphosis, he made an important discovery. Molluscs typically hatch from their eggs as trochophore larvae – the wheel-like ciliated larvae we met in chapter six – but Darwin observed that barnacles metamorphose through nauplius larvae, which are only found in a single phylum – the crustaceans, which include shrimps, crabs, and lobsters. From this Darwin correctly deduced that barnacles were not molluscs but crustaceans, as 'a glance at the larvae shows this to be the case in an unmistakable manner'.

Yet when, in his book, Williamson looked in some detail at the life cycles of the crustaceans, the extrapolation of the larval types to their corresponding adults did not always work true. Indeed, as Williamson deepened his investigation, such an extrapolation to the tree of life seemed to be very misleading. He considered the relationships between the biggest of the evolutionary divisions, the phyla, and declared that the orthodox tree of life, despite its undeniable usefulness, needed to be considered anew. 'The reader should be warned that, if my views are accepted, little but the stump of this tree will remain intact'.

He next turned to the specifics of evolutionary development and its implications for metamorphosis. Some features of echinoderm metamorphosis remain much the same, regardless of the type of larva. For instance, paired coelomic sacs form inside the body. One of these, usually on the left, grows bigger than the others, and it is here that the next step in the animal's metamorphosis begins. In the wall of that sac, pluripotent cells – the equivalent of stem cells – grow into the juvenile adult. This organism develops five lobes, which become radially arranged. From this strange start, the form of the juvenile adult echinoderm emerges. In many cases, the embryonic development really does amount to a new beginning, a novel organism developing according to a new blueprint to construct the adult animal. In echinoderms, the pentaradial symmetry owes nothing to the larva, and the bulk of the adult structures, including the arms of the adult starfish or brittle star, develop independently of any larval arm or lobe. Indeed, the adult starfish or brittle star appears to originate and grow within the larva like a separate and independent parasitic being.

Most astonishing, in the development of both starfish and sea urchins, the mouth of the adult arises as an entirely new structure unrelated in any way to the mouth of the larva. When one considers the importance of the evolution of the mouth in developmental biology – its position and role determining the great division between protostomy, where the first dimple, or blastopore, becomes the mouth, and deuterostomy, where the blastopore becomes the anus – it really is difficult to imagine how or why such a vital structure could be abandoned as part of a linear descent-with-modification trajectory. In particular, Williamson highlighted the fact that, although the larvae are clearly deuterostome in their developmental patterns, the term is actually meaningless when applied to the adults, since the mouth of the adult is no longer derived from the blastopore of the embryo.

We have seen how the role of the special pluripotent cells lining the coelomic sac is similar to the role of the imaginal discs during the catastrophic metamorphic change and reprogramming in the insect pupa. So striking is this commonality, it makes one wonder if there could also be some evolutionary, or developmental, commonality.

In both insect and marine species that metamorphose through the activation of stem cells, with the accompanying cataclysmic changes, there appeared to be two separate developmental blueprints involved. For most biologists this was explained through the separate phases of evolutionary adaptation for each phase of metamorphosis within the same organism. But for Williamson, the more likely explanation lay with the amalgamation, through hybridisation, of two separate evolutionary lineages.

In chapter after chapter of his book, Williamson outlined his reasons for proposing that the entire echinoderm phylum had evolved from directly developing ancestors that were radially symmetrical throughout their lives. Then he went on to note similar anomalies between larval and adult evolution in many other phyla, including the familiar molluscs. In these various marine phyla, the adults were radically different from each other in form and life cycles, yet some members of each of the phyla hatched as wheel-like trochophore larvae. The trochophore is as unique in its appearance and structure as the easel-shaped pluteus of the echinoderms. Because of their common trochophore larvae, these different phyla, with their radically different adult animals, are assumed to be offshoots of the same branch of the evolutionary tree. But Williamson questioned what the bottom-crawling, multi-bristled polychaete worm *Nereis* could possibly share with the abalone molluscs, decked out in their beautiful spiral shells and crawling over the rock bottoms on a single powerful foot. To build his evidence, Williamson included illustrations of metamorphic development among these common trocophores that defy verbal description. For example, the metamorphosis of the polychaete worm *Owenia fusiformis* is more spectacularly bizarre than Ridley Scott's *Alien*.

In his view, a close evolutionary link between these very different phyla was unlikely, and as Williamson states, 'few of those who have favoured convergence as the explanation of the similarities have given any evidence for their views'. It would, indeed, be hard to do so since the huge evolutionary differences between the phyla were already apparent at the beginning of the Cambrian period, some 570 million years ago. Moreover, unlike the very different adult developments, the development from fertilised egg to trochophore larvae in these phyla is very similar, suggesting that the common larval shapes really do derive from a common evolutionary origin and not from convergence. He challenged traditional evolutionary biologists to explain the minimal differences between larvae compared to the remarkable differences between adults in the same animals whose life cycles are purported to have existed for hundreds of millions of years.

Whatever one's view of Williamson's theory, it's clear that he possesses a formidable knowledge of marine invertebrate larvae and he makes a good argument. There would appear to be glaring anomalies in the evolutionary relationships of marine larvae and adults. Yet Williamson himself had to admit that many questions remained to be answered in his own theory. How could he distinguish a larval form that had, in his opinion, been transferred from one taxon to another from a form that had evolved in a gradualist and linear manner? Did the transfer, if and when it happened, result from an isolated hybridisation between individuals of different species or did it represent repeated crosses between different groups? These were not questions to be easily answered. From a detached perspective, they highlighted a significant omission running through all of his theorising: he had no knowledge of what was happening at the genetic level. More than anything, the lack of genetic input had damaged his credibility in the eyes of the colleagues at his American lectures. He knew this had to be rectified. Another difficulty lay in the fact that, in 1992, hybridisation was an unfashionable subject for budding evolutionary biologists. The genetic implications of the union of two different genomes, with all of their complex dynamics and infrastructure, were daunting. Yet hybrid experiments appeared to work – and hybrid organisms certainly existed. How would a hybrid history affect the developmental pathways of an organism? Nobody knew. Without getting a handle on the genetics, the potential implications of hybridisation as an evolutionary force would remain unknown and this, in turn, would continue to undermine Williamson's theory. The problem remained that Don Williamson was no more a geneticist than Charles Darwin had been. He needed assistance.

Then, in February 1994, Williamson received a letter from Michael Hart, a postdoctoral fellow working at the Institute of Molecular Biology and Biochemistry at the Simon Fraser University, British Columbia. Hart enquired about Williamson's 1990 sea urchin experiment. He informed Williamson that he was a geneticist by training, and he was offering to look at the genetics of the hybrid offspring – if some were still alive.

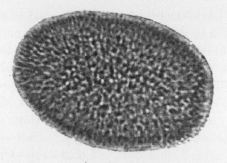

*Planula larva of jellyfish*

## 17

# The First Genetic Testing

MICHAEL HART HAD READ Williamson's book in 1992 when he was a graduate student working with Richard Strathmann, Professor of Zoology at the University of Washington. Strathmann is a distinguished marine biologist and Director of the Friday Harbor Laboratories, located on San Juan Island, off the Pacific coast of Washington State, whose stated research interest is 'why the beautiful and diverse patterns of development have evolved as they have instead of in other ways'.[1] Hart and Strathmann found Williamson's *Larvae and Evolution* provocative, though they remained extremely sceptical about its content. Strathmann had, moreover, written a thoughtful if critical review of the book for the *Quarterly Review of Biology*.[2] Even today the review reads as determinedly open-minded while encapsulating the disbelief shown by many scientists towards Williamson's iconoclastic ideas. In the first paragraph, Strathmann writes, 'If the hypotheses and experimental results in this book are correct, then much of evolutionary theory, systematics, and developmental biology will need revision'. Williamson, in Strathmann's opinion, only needed to achieve plausibility, not concrete proof, to cast doubt on some of evolutionary biology's fundamental assumptions. To achieve plausibility, Strathmann went on, Williamson's theory needed to provide convincing answers to three separate questions.

First, did the larval-adult incongruities Williamson highlighted merit his claim for a radically different evolutionary process, such as hybridisation, or could they be accounted for by orthodox theory? In Strathmann's opinion, convergent evolution could produce incongruities between larval and adult developments. Even the cataclysmic changes seen in echinoderm metamorphosis could be interpreted as an adaptation, one that allowed for rapid conversion from one developmental form and habitat to another. Second, were the developmental consequences of Williamson's purported hybridisations consistent with the prevailing evidence of how hybridisation affects normal developmental processes? Strathmann did not think so, but he acknowledged that hybridisation and its genomic implications were specialist territory and would require a developmental biologist to present a more detailed critique. At the end of his book, Williamson proposed a number of biochemical, immunogical, and genetic tests that would, in his opinion, confirm that his hybrids were genuine. These were the basis of Strathmann's third line of inquiry: were the proposed tests going to be sufficient? Strathmann doubted it. Williamson had not focused on genes important for structural development, so none of these tests would be capable of confirming his more general hypothesis that traits of adults and larvae from different taxonomic groups had been combined into one life history by transfer of genetic material.

Strathmann looked at Williamson's hypothesis as tantamount to believing in miracles, and he quoted the philosopher
David Hume's line: 'Miracles are by definition highly improbable events'. Yet in spite of his doubts, and in spite of what he
judged to be the manifold weaknesses of Williamson's hypothesis, Strathmann considered the publication of Williamson's
book to have been appropriate: 'If a revolutionary hypothesis
is suppressed, it cannot be evaluated'. After all, genetic transfers across the species barrier were known to occur, for example through symbiotic merging of distantly related life forms.
And even small-scale genetic changes might have major effects
on body growth and form. Strathmann perceived a significant
implication arising from Williamson's experiments. 'Could
transfers of genetic material between distantly related organisms have important morphogenetic [developmental] consequences?' He could not offer an example of such a scenario '…
but attempts to construct one could succeed or fail instructively,
and much hangs on the answer'.

Where so many biologists and editors had dismissed Williamson or, worse, attempted to prevent the publication of
his larval transfer theory, a more open mind saw beyond his
experiments, and even beyond his theory, to an opportunity
for greater scientific enlightenment. In an important sense, and
this is the sense in which Strathmann saw it, Williamson's ideas
and experiments would advance the cause of science, whether
they stood or they failed. But what more did Williamson have
to do to convince the world of biology to take him seriously?
Strathmann was in no doubt: 'Positive results from hybridisations of distantly related animals would directly confirm Williamson's proposed basis for adult-larval incongruities'.

To Strathmann's credit, he attempted to repeat Williamson's experiment with species locally available in the San Juan
archipelago. However sceptically he referred to his own cross-
phyletic hybridisation experiments as 'attempts at cold fusion',

he nevertheless encouraged other biologists to do the same. 'Why not', he urged, 'when it is so easy?' This is exactly what Williamson had been asking his colleagues to do for years. Strathmann had some of his students try to cross urochordates, also known as tunicates, with echinoderms, Hart later told me – but these experiments failed to produce embryos. Hart also confessed that he was just as sceptical as Strathmann when he read Williamson's book in 1992, '… but we both thought it was worth pursuing or at least thinking about'.[3] This had led him to contact Williamson in search of viable genetic material from the hybridisation experiment.

We might recall that in this experiment, conducted in 1990, Williamson had crossed the eggs of the sea squirt *Ascidia mentula* with sperm of the edible sea urchin, *Echinus esculentus*. Out of twenty-five attempts at cross-phlyletic hybridisation, the great majority had produced no larvae at all. In six attempts, eggs had hatched as free-swimming forms that resembled ciliated blastulae. In three experiments, larvae had developed beyond the gastrula stage, producing the paternally-derived pluteus larvae in various numbers. About seventy-five per cent of the hybrid plutei were stunted or showed some deformity, for example an arm failing to develop. Similar stunting or deformity of larvae had been witnessed by other biologists conducting hybridisation experiments. About twenty-five per cent of the presumed hybrid larvae showed no deformities and seemed indistinguishable from normal sea urchin larvae. But none of these metamorphosed. Williamson repeated the experiment, this time producing larvae that did metamorphose. Indeed, the new experiment produced over three thousand pluteus larvae, which Williamson cultured further, supplying them with diatoms, a type of algae, as food. More than seventy of these larvae developed juvenile rudiments (the first stage of adult development) inside their coelomic sacs, and twenty successfully metamorphosed over the subsequent thirty-seven to fifty days. These were placed in a small aquarium tank, fed, and monitored. A year after first fertilisation, four of these survived as apparently healthy sea urchins. Williamson hoped that these would reach maturity so he could breed from them, crossing one hybrid urchin with another (if he obtained different sexes) and crossing hybrid urchins with non-hybrid urchins and also with ascidians.

Of the pluteus larvae that had resulted from this last experiment, several hundred attained full larval development, but without producing any trace of juvenile development inside them. These continued to swim actively for a time before gradually resorbing their pluteal arms and condensing into a spherical ball of cells with a large rounded protuberance at one side. From this protuberance, these "spheroids" could fix themselves to the walls of the rearing bowl with thread-like organs of attachment. They were seen to release themselves, swim away, and reattach in similar fashion elsewhere. In Williamson's words, 'Although they developed from pluteus larvae, they are quite unlike any known stage in the development of an echinoderm'. He transferred the spheroids to a small aquarium at fifteen degrees Celsius and waited to see whether they developed further; however, they all died over time, without further evidence of development.

Commenting on this experiment in his book, Williamson freely admitted that there was no evidence of sea squirts and sea urchins hybridising in nature. Crossing them experimentally did not prove that an ancestor of any animal group ever acquired a larval form by cross-fertilisation.[4] But it convinced him that he was at least on the right track: 'The experiment shows, however, that the first step in the process of transferring a larval form from one species to a very distantly related one by cross-fertilisation is entirely feasible'.

The geneticist Hart was sceptical of such cross-fertilisation because of his knowledge of developmental biology. To his thinking, Williamson's description suggested there was a wholesale takeover of the maternal developmental programme by the paternal genome:

> *The reason I found his example involving tunicate eggs and sea urchin sperm so remarkable – and unexpected, and needing corroboration from other evidence – is that tunicate eggs and embryos have a highly deterministic developmental program. Early [division of the egg], movement of cells, and the fate of these cells, happen without the [fertilised egg's] expression of either the maternal or paternal part of [its] genome. It is all run by maternal messenger RNA plus interactions among the cells of the early embryo. So it is not clear how a sea urchin sperm nucleus could provoke a tunicate egg to become a sea urchin embryo and larva.[5]*

It was unfortunate that by the time Hart contacted Williamson in 1994, the plutei and spheroids had all died off. These had been by far the commonest larval offspring and would have been the ideal source for genetic investigation. Williamson had saved plenty of tissue material from the hybrids, but it was preserved in formalin, not alcohol; he was unaware that this would make them unsuitable for DNA analysis. However, three of the urchins that had developed from the paternal pattern of pluteus larvae were still alive. He and his wife Enid were about to take a holiday, again visiting his sister Mary in Yorkshire. He tweaked tube feet from the living hybrid urchins, cooled them in a fridge, put them in a Thermos flask, and took them with him. Tube feet will remain viable in these circumstances for at least a week. From Thirsk, he travelled twenty miles south to visit Tom Crompton, a research student at York University. Crompton froze the specimens at minus seventy degrees Celsius and sent them in an insulated container to Michael Hart in the United States.

On receiving the samples, Hart formulated his experimental method to examine their DNA content. There are two specific sites in any animal that contain distinctively different DNA. The great majority of DNA is found in the nuclear chromosomes. This is the common inheritance of the nuclear genes from both the maternal and paternal line. The other site is in mitochondria – tiny power-producing packets that reside outside of the cell's nucleus. Mitochondria were once free-living, oxygen-breathing bacteria that were symbiotically incorporated into the nucleated cells of the forerunners of all plants and animals, and they still contain a residuum of bacterial-type DNA. Mitochondria are inherited solely through the eggs of the maternal line in mammals, but the situation in sea urchins appears to be more complex – with some reports of sperm contribution to the mitochondrial DNA of the offspring.

In Williamson's experiment, the maternal line was the sea squirt *Ascidia mentula* and the paternal line was the sea urchin *Echinus esculentus*, so Hart set out to investigate the genetics in two ways. He used the polymerase chain reaction (PCR) to amplify a portion of one of the mitochondrial genes, which should be maternally inherited, and he compared the nucleotide coding of this fragment with known DNA sequences of the same gene in sea squirts and sea urchins. This gene was known to be highly conserved, meaning it was vital to genetic function and relatively stable within the species. Then he used PCR to amplify a portion of the nuclear 28s ribosomal RNA gene of both species, again a gene known to be highly conserved.

The mitochondrial DNA sequences of the putative hybrids were almost identical to those of the sea urchin *Echinus esculentus* but showed sixty-three nucleotide differences with the tunicate *Ascidia mentula*. The nuclear gene comparisons were technically more difficult to compare, but Hart was convinced this also fitted with the known sequences of sea urchins. Overall he could find no sequences typical of the sea squirt. Hart concluded: 'These results indicate that the putative hybrid developed from a sea urchin egg fertilised by a sea urchin sperm, and not from a tunicate egg as described by Williamson'.

Hart's genetic analysis led to a difference of opinion between the two biologists. Hart did not suggest that Williamson had attempted any deception. Williamson had freely provided him with the tissue samples and had encouraged the testing of his hypothesis. Even though the results had contradicted his expectations, Hart still commended Williamson's willingness to 'hazard an unpopular idea' and 'his good humour in response to scepticism'. Indeed, Hart now produced a theory of his own to explain Williamson's findings. Some sea urchins were known to be hermaphrodites. They produced both eggs and sperm and were capable of self-fertilising their own eggs. If the sperm collected from the sea urchins had been contaminated with self-fertilised eggs of this nature, it would explain those offspring that had gone on to form pluteus larvae and metamorphose to sea urchins.

Williamson refuted this hermaphrodite theory. Given the large numbers of plutei larvae, Hart's hypothesis suggested that most if not all of the eggs used were sea urchin rather than tunicate, which the experienced marine biologist rejected. Williamson noted that the eggs of the two different phyla are visually distinct – sea urchin eggs are surrounded by a layer of jelly, whereas sea squirt eggs are surrounded by spherical follicle cells – and he had taken considerable care to exclude hermaphroditic sea urchin eggs during his experiments. Moreover, Dr Hilary B. Moore had studied the population of *Echinus esculentus* on the Port Erin breakwater in the 1930s, and he found that only one in three thousand was hermaphrodite.

Perhaps critical to the disagreement between Hart and Williamson was the fact that the urchins that provided the genetic material came from only a small minority of the pluteus larvae. Over ninety per cent of the plutei in the same culture had retracted their arms to become spheroids, a form unknown in the natural development of sea urchins. The great majority of the hybrid offspring, which had hatched to the paternal pattern of plutei larvae before developing to spheroids, had not been genetically examined.

Despite Williamson's protests, Hart went on to publish the results of his genetic examination under his name alone, in 1996.[6] In the publication he described Williamson's idea as an 'improbable hypothesis [that] implies that the evolutionary history of the major [evolutionary groups] is a ... web of hybridisations and divergences'. In describing why he had undertaken the experiment, he wrote: 'Williamson's heretical view ... was widely dismissed, though some have advocated testing his experimental results.' When I later wrote to Hart, I asked him if such language suggested a bias against Williamson's theory at the outset. He replied promptly, and in some detail. He had conducted his examination while he was an unsecured postdoctoral fellow on a two-year appointment. He had not been biased against Williamson or his hypothesis. Indeed, genetic analysis of the hybrids had seemed a wonderful opportunity for someone looking to secure a permanent academic appointment. 'It would have been an enormous boost to my career (not to mention Don's hypothesis) if I had found tunicate DNA in his hybrid sample. Far from being disposed against his ideas, I was in fact rather disappointed at the bland predictability of my results'.

Whatever their differences, both biologists are undoubtedly sincere and dedicated in their respective approaches to scientific truth. But Hart's conclusions were a crushing blow to Williamson's hopes of garnering a genetic confirmation for his results. In the introduction to his paper, Hart wrote, 'Testing improbable or heretical hypotheses ensures that such hypotheses succeed or fail on their merits and not on their relation to orthodoxy'. The implication, as far as Hart was concerned, was evident. Williamson's hypothesis had failed to stand up to rigorous investigation. It should be dismissed.

*Hypothetical dipleurula larva (gut outlined)*

## 18

# How Novelties Can Arise

RUDOLPH A. RAFF IS Professor of Biology and Director of the Institute for Molecular and Cellular Biology at Indiana University. He has long been interested in the metamorphosis of sea urchins. Hybrids of different species, genera, and even families of sea urchins are known to undergo successful metamorphosis. Raff is particularly fascinated with the elimination of the larval stage in certain species of these marine invertebrates, a process known as direct development, believing that it offers a window onto the study of evolutionary modification. In a 1987 paper, he observed that direct development appears to have evolved independently several times, and in several different ways.[1] Over decades, Raff and his colleagues have studied the effects of hybridisation between two species within the same genus of sea urchins, looking in particular at the effects this has on the larval phase of the hybrid offspring.

The purple sea urchin, *Heliocidaris erythrogramma*, is the most prevalent urchin found around the south Australian coast, ranging from Port Stephens in New South Wales in the east to Shark Bay in Western Australia, where it carves out hemispherical hollows in the rocks at the lowest tide levels. Colours can vary dramatically, from a light olive green, through pink, to dark purple, causing previous generations of biologists to mistake the purple sea urchin for three separate species.

The brown sea urchin, *Heliocidaris tuberculata*, is also a common Australian resident, found in a wide distribution on the east coast, from southern Queensland to mid New South Wales. Its main distinguishing features are the bright orange colour of its spines and, in contrast with the purple sea urchin, the bluntness of its spines at the tip. The brown sea urchin develops through the normal metamorphic phase of a pluteus larva, which feeds and grows in the surface waters for six weeks or so before metamorphosing to a juvenile adult; the purple sea urchin, in contrast, undergoes an interim pattern of direct development, in which the larva has been simplified to a giant – relatively speaking – potato-shaped larval phase, devoid of mouth, gut, arms, or lobes, before metamorphosing into the juvenile after just four days.

Studies suggest that the two species diverged from a common ancestor about ten million years ago, and some time thereafter the purple sea urchin larva lost its feeding structures, even its gut. The purple sea urchin's life cycle has gained many new traits in the meantime, including changes in embryo development, notably in the early cleavage pattern of its ball of cells, and a novel mechanism of gastrula formation – the stage where the embryo has developed into a hollow ball with a primitive gut.[2] This has been accompanied by a major increase in yolk, essential to the purple sea urchin larva's non-feeding existence. There have been so many other complex adaptations of the two species, including evolutionary modification of the animal's sperm, that if an observer, ignorant of the adult developments, were to compare and contrast the present-day pluteus larva of the brown sea urchin with the fattened, yolk-rich, and relatively featureless larva of the purple sea urchin, they would conclude they were phases in the lifecycle of different life forms.[3]

Raff and his colleagues have developed considerable expertise in hybridising the two species of urchins, and, in a landmark series of experiments, they have studied the developmental outcome. Hybridisation has proved remarkably successful when the purple sea urchin provides the eggs. But reverse crosses, involving the eggs of the brown sea urchin fertilised by purple sea urchin sperm, had to be abandoned after they were found to be non-viable. Using state-of-the-art assessment technology, Raff and his team have worked out some of the complex accommodations that occur at genetic and molecular levels within the hybrids. It has proved to be an enlightening enterprise.

Past prejudices dismissed hybrid animal offspring as inferior to those of "pure" or "naturally breeding" parents. Sometimes this is the case – but it is by no means an absolute rule. For instance, when biologists investigated the outcome of hybrid crosses between two different species of sea urchins within the *Echinometra* genus that live in the coastal waters off Okinawa, they observed high rates of fertilisation, and the resulting hybrids were vigorous in their growth and metamorphosed to fully reproductive adults.[4] However, the investigating scientists found little evidence for frequent hybridisation in their oceanic habitat. They could offer no obvious explanation for this, since the two species shared the same ecology and their breeding times overlapped. Elsewhere the same researchers found abundant evidence of hybridisation between other species of sea urchins in their natural habitats.[5]

When Williamson first presented his larval transfer theory, the great majority of evolutionary biologists were sceptical of any role for hybridisation in animal evolution, assuming that hybridisation led to genetic instability and infertile offspring. But biologists today are becoming increasingly aware of the importance of hybridisation in certain cases of animal evolution. The various hybrids born of these interspecies crosses are even described as "robust", indicating that the two parental genomes, whatever their evolved differences, appear to have united in a seemingly healthy way in the hybrid genome. Closer study of such hybrid offspring has revealed many surprises.

Raff knows that in their development, hybrid echinoids tend to follow the maternal features up to and including the point where the gastrula is formed, after which they tend to exhibit paternal features. This observation may have relevance to Williamson's findings of paternal plutei patterns of larvae in his sea urchin–sea squirt crosses, since it suggests that genomes may contain in-built organisational switches that trigger different developmental roles at various stages of the developmental process. Further, as Raff's hybrids switched from a maternal to a paternal pattern of development, his sea urchin hybrids acquired pluteus feeding structures – which had been lost to the maternal line for millions of years.

What Raff and his colleagues observed was far more complex than merely a reversion to the paternal pluteus: the hybrid larvae developed along lines distinct from either parental species.[6] The initial cleavage of the embryo followed the maternal purple sea urchin pattern up to and including the formation of the gastrula. Under normal circumstances, at this point a larva following the maternal developmental pattern would have adopted the relatively featureless shape of the purple sea urchin larva, devoid of arms or a gut and with only a single, incomplete ciliary band. Instead, over days two to three, the hybrid larva transformed to a deeply lobed, flattened form with distinct mouth and anus separated by a single highly convoluted ciliary band. Raff suggests this could be a most unusual larval development.

To make sense of it, we might recall the basal branching of the orthodox tree of life, where the first dimple opening in the developing embryo, the blastopore, becomes either the future mouth, as in the protostomes, or the future anus, as in the deuterostomes. It has long been held that there are two basic types of primary invertebrate larvae, corresponding to those same basal branches. The primary larva of the protostomes is assumed to be the familiar trochophore, the wheel-like larva of many species of molluscs, while the primary larva of the deuterostomes is presumed to have modified and changed over the vastness of evolutionary time. But developmental biologists had formulated a hypothetical primary larva of the deuterostomes, based on deuterostome embryonic development and on commonalities among various deuterostome larvae – a hitherto unseen larva they called a "dipleurula". According to the developmental biologist Pat Willmer, 'the dipleurula does not strictly exist as a larval form, and may never have done; it is merely a [theoretical] archetype'.[7] Now, if Raff and his colleagues were correct, it appeared they had, however fleetingly, resurrected this archetype as a developmental phase in a real organism.

Moreover, as the hybrid larvae developed and changed, they exhibited a range of "ancestral" features, some of which resembled the bipinnaria larvae of the starfish and others the auricularia larvae of the sea cucumber. They observed how the hybrid larvae continued to change, until, by day five, they resembled the paternal plutei. But Raff and his co-authors took pains to point out that these later-stage hybrid larvae did not represent a reversion to a pluteus. Indeed, in many important respects the hybrids represented neither a paternal nor a maternal larva. Their opinion was emphatic: 'We conclude that the developmental mode is truly a "hybrid" one, in which features from both parents combine to generate a new pathway'.

After seven days the hybrid larvae metamorphosed to juvenile adults. Although the development of the juvenile was often imperfect, the biologists concluded that the hybrids revealed '... an amazing and unexpected potential for integration of the disparate parental pathways', shedding new light on ways in which embryonic and larval development can be radically reorganised by hybrid crosses while at the same time allowing a coherent development. Unfortunately, the resulting young adults did not live long enough for the biologists to be able to perform cross-hybrid crosses or back crosses with either of the parental species.[8]

To the scientists, the unique structural and genetic pathways revealed by the hybrids suggested that this '... might be a way in which developmental novelties can arise'. And although they were too cautious to say so, it also suggests that major evolutionary change might come about through hybridisation.

The researches undertaken by the Indiana group are important in another respect. When Mark G. Nielsen and his colleagues examined the genetics and molecular biology of what was happening in these hybrid experiments, comparing the hybrid offspring with the parental species, they found higher levels of paternal gene expression in the development of the hybrid embryos than have been reported from other hybrid sea urchins.[9] To put it simply, the development of the hybrid embryos were as dependent on the paternal genetic inheritance as on the maternal. Moreover the genome appeared to choose paternal or maternal genetic inheritance in a non-random fashion, as if it could recognise the parent of origin and would choose paternal for certain developmental pathways and maternal for other ones. The researchers also found a great deal of genetic "creativity" in the hybrids, including new sites of genetic expression, genes with new additive expression, and novel regulatory gene interactions.

This creativity has developmental implications at the cellular and organisational levels. In sea urchins, as in insects, the cells of the skin layer, or ectoderm, play a crucial role in body form and development. The purple sea urchin and the brown sea urchin exhibit very different ectodermal cell types, but in the hybrid offspring these types combined into an intermediate form. Extraordinary as it might seem, the merger of the two parental genomes produced, in one fell swoop, a new biological entity – with a genomic developmental pathway distinct from that contained in either of its parents.

Nielsen, Raff, and their colleagues are cautious in extrapolating these findings to a major evolutionary role for hybridisation because, in nature, hybrid offspring – and a potential new species – might or might not survive. Two factors might help determine survival or extinction: whether the offspring can compete effectively with either parental species for resources within the same ecosystem and whether it might possess some new properties that enabled it to colonise ecological niches not habitable by the parents. We also need to consider the likelihood, as suggested by Williamson as part of his larval transfer theory, of repeated hybridisation events between the same two parental species over the vastness of evolutionary time. Repeated crosses, in different seasons, or in varying ecologies, would make it rather more likely that, from all such trial and experiment, hybrid crosses would ultimately find success.

In 2002, Professor Meiko Komatsu and his graduate student, Takako Chimura, of Toyama University, Japan, reported interspecies and interclass hybrid studies involving the sand dollar *Peronella japonica*, which metamorphoses through a non-feeding, two-armed pluteus, and four different species of starfish from two separate genera.[10] The various permutations of male and female cross-fertilisations resulted in six successful hybrid pairings, some involving cross-species fertilisations and some apparently cross-class. In every crop of successful hybrid offspring, some metamorphosed to juveniles. This was the first demonstration that cross-class fertilisations involving these marine organisms could result in normal juveniles. Komatsu and Chimura were sufficiently encouraged to undertake further studies.[11]

As its Latin name, for "handsome", suggests, the sea urchin *Hemicentrotus pulcherrimus* is a noble creature. It shares the coastal waters of Japan with a large yellow-and-purple starfish known as the Northern Pacific sea star, *Asterias amurensis*. Komatsu and Chimura conducted hybrid crosses between these two classes of echinoderms, with the sea urchin providing the eggs.[12] These crosses resulted in offspring that developed to pluteus larvae, which in turn metamorphosed to juveniles. But in this case, genetic analysis suggested the larvae and the juveniles were not the result of true crosses but had developed from the maternal egg alone, a single-parent, or parthenogenetic development.

Nevertheless, after decades of scepticism, these observations, together with growing evidence from many other scientific departments worldwide, are confirming the importance of hybridisation as a mechanism for hereditary change in evolution.[13]

*Budding hybrid spheroid*

## 19

# A New Life Form

IN 1996, AN ASPIRING marine biologist, Sebastian Holmes, happened to share an office with Don Williamson in the Port Erin Marine Laboratory. Holmes had recently graduated in biology at Bangor University, where he had become interested in the interactions between molluscs, such as limpets and periwinkles, and the other life forms of the oceanside community. Now, for his PhD, he had come to the Isle of Man to study barnacle populations. During the course of their daily contact, Williamson explained the nature of his larval transfer theory. As a result of his stroke, Williamson had been unable to continue practical experimentation, but he had not abandoned his theory. 'At the time, I thought, "That's really smart. I like that",' Holmes later recalled of the theory. 'And I just logged it away in my mind.'[1]

Where Williamson's interest was in larval evolution, Holmes was intrigued by the potential of hybridisation to form

completely new animals. It seemed to him that if hybridisation was a genuine force in nature, it would give rise to sudden, large-scale genetic novelty, speeding up the evolutionary
process. The hybridisation experiments conducted by Rudolph
Raff and his colleagues seemed to confirm exactly that – and
they were only crossing related species of sea urchins within the
same genus. Hybridising animals from different orders, classes,
and possibly even phyla, if it proved fruitful, would take this
idea a major step further. In the years that followed, Holmes
kept in touch with Williamson, hoping that at some stage they
might have the opportunity to work together.

Holmes completed his PhD in 1998 and took a post-doctoral
fellowship at the Netherlands Institute for Sea Research, where
he extended his investigations to marine population genetics. A
few years later, while waiting for consideration of further fund-
ing from the Dutch National Environmental Research Council, he thought again about Williamson and his extraordinary
theory. It occurred to Holmes that he was now in a position
to help Williamson repeat his hybridisation experiments. All
they needed was financial support. I happened to mention this
in passing to Lynn Margulis, who had first alerted me to Williamson's work. She generously offered Holmes and Williamson some of the prize money she had recently been awarded
for scientific achievement.[2] After the long and frustrating gap
enforced by Williamson's disability, the two marine scientists
were ready to make plans for a new experiment.

'What proof would be needed from the experiment to convince the world of science that Williamson's theory should be
taken seriously?' I asked Holmes at the time.

'My take on Don's work,' he replied, 'is that he has actually
done it already. He has shown that the cross-phyletic hybridisation works. I would repeat his experiment, backing it up with
some molecular biology. I think that once you get two different phyla to cross and show that the offspring survive, that's it.

For me that would be sufficient proof. I would also think that for the general scientific community that would be enough. Others could investigate it further. People could go about fine tuning it.'[3]

In September 2002, Holmes, accompanied by a biology student named Nic Boerboom, arrived in Port Erin aiming to repeat Williamson's 1990 experiments. They brought some useful equipment with them from Holland, including a photomicrography computer setup that would enable them to capture microscopic images of any putative hybrid offspring and their development. Throughout September and October, they collected mature specimens of the green sea urchin, *Psammechinus miliaris*, by hand from Gansey Point, Port St Mary; they collected specimens of another sea urchin, the so-called "seapotato", or *Echinocardium cordatum*, from Derby Haven; they harvested from sand at a depth of a few metres in Port Erin Bay the common starfish, *Asterias rubens*, the familiar orange species found in almost every marine ecosystem around the British Isles; and they collected the sea squirts *Ascidiella aspersa* and *Ciona intestinalis* from the seawater storage tank back at the laboratory. To complete their preparations, they trawled from a boat at various locations off the island to collect the "bloody Henry" starfish, *Henricia oculata*, and the spiny starfish, *Marthasterias glacialis*. All these animals were conveyed to the lab, where they were washed with filtered seawater and maintained in aerated seawater that had also been filtered. Only now did Holmes, Boerboom, Williamson, and Williamson's wife, Enid, combine forces to begin harvesting the eggs and sperm from this array of animals.

Seven reciprocal crosses were attempted, making fourteen hybridisation experiments in total. Control experiments were set up against all of the potential hybrids. Ten involved crosses between different phyla and four between different orders. The labelled eggs were placed in a beaker of seawater and just sufficient of the appropriate sperm were added to tinge the water milky. After twenty minutes, the excess sperm was washed from the now fertilised eggs, which were split into separate containers and allowed to develop. Any eggs that showed no signs of developing were discarded.

Ten of the fourteen crosses, consisting of five reciprocal groups, resulted in successful fertilisations. In two of the five reciprocal groups, one at cross-phyletic level and one at cross-order level, cell division halted before the stage of a blastula and the cells disintegrated. Of the remaining three successful fertilisations, all at cross-phyletic level, two proceeded to the development of larvae, with the larvae adopting the forms expected of the echinoderm (starfish or sea urchin) parent as opposed to the sea squirt parent, regardless of whether the echinoderm was the maternal or paternal partner. But in the majority of these larval offspring, development also arrested, usually after between seventy-two and ninety-six hours, causing the larvae to disintegrate.

The researchers obtained long-term offspring survivors in a single cross-phyletic hybrid, involving eggs from the sea urchin *Psammechinus miliaris* and sperm from the sea squirt *Ascidiella aspersa*. In this trial, development proceeded to the formation of an early four-armed pluteus larva – in other words, it followed the maternal echinoderm pattern. This was a similar cross-phyletic grouping to the one used in Williamson's earlier experiments, but it reversed the parentage of the two contributing species.

In this more successful cross, the sea squirt sperm had been pipetted from a visible cloud in seawater emanating from an isolated individual – this meant the sperm was less concentrated than in the earlier experiments, closer to the concentration that would normally be found during sexual reproduction in the ocean. The three scientists followed the methodology established by Raff, washing the sea urchin eggs in acidic seawater for forty-five seconds, rinsing them, and then fertilising them with ascidian sperm. Controls were also set up, involving normal fertilisations of *Psammechinus miliaris* eggs with *Psammechinus miliaris* sperm. These developed into eight-armed plutei, the normal larva of the parental species.

The great majority of these hybrid eggs divided, hatched as ciliated blastulas, and developed to four-armed plutei two weeks after hatching. During the third week, however, the hybrid plutei retracted their four arms and became spheroids. These soon settled to the bottom of the culture plates, but they lacked the organs of attachment seen in the spheroids during the 1990 hybridisations. The new spheroids were about a fifth to a tenth of a millimetre in diameter, and, lacking cilia, they could move at no more than one millimetre per hour over the bottom of the dish. Now and then, in dramatic fashion, they reproduced, the body elongating, developing a constricted ring in the middle, and dividing through a visible, and frequently photographed, budding into two free-living individuals. Many spheroids produced clumps of cells of up to two millimetres in diameter, displaying what appeared to be differentiation into distinct tissues. Indeed, three of these went on to resemble an early stage in the development of a juvenile ascidian, although they failed to develop into adults. Over hours, or days, they became indistinguishable from the hundreds of other structureless clumps of cells.[4] Meanwhile, the spheroids went on living and reproducing by budding, for many months.

Although the results were fascinating, Williamson was left somewhat disappointed. He had hoped the hybrids would go on to produce viable and fecund adults closer in complexity to the parental generations. But the success of the cross-phyletic fertilisations and the reproduction, through budding, of the new spheroids remained immensely curious.

Holmes had stated before conducting the experiment: 'I

think that once you get two different phyla to cross and show that the offspring survive, that's it. For me, that would be sufficient proof.' And Richard Strathmann at the University of Washington had also suggested that Williamson only needed to achieve plausibility, not concrete proof.

In 2003, Holmes sent me more than a hundred photomicrographic images of the hybrid offspring, taken at each stage in their development from fertilised eggs, through plutei and then spheroids. I gazed, astonished, at images that captured the spheroids budding. Larvae do not reproduce. Budding is a form of asexual reproduction that is seen in colonial marine invertebrates, such as corals and moss animals. It is not seen in sea urchins, such as *Psammechinus miliaris*, which had provided the eggs for the hybrid cross. And while budding is a feature of some colonial forms of sea squirts, it is unknown in *Ascidiella aspersa*, which had provided the sperm.

The complex genomic reorganisations that follow hybridisation are only poorly understood at present. This latest cross-phyletic experiment, in merging two radically different life forms, was, to put it mildly, adventurous. The sea squirt, with its tadpole larva, has a tenuous link to the vertebrates, while the sea urchin belongs to a phylum that is widely separated from that of the sea squirts, both on the tree of life and by separate evolutionary trajectories over hundreds of millions of years. If Williamson is right in his larval transfer theory, the larval forms of sea squirts, and thus the metamorphic life cycle, arose hundreds of millions of years ago from a hybrid fusion of two dissimilar species. But this is presumed by Williamson to have taken place at a time when animals were at a purportedly more malleable stage of their evolution, before natural selection had honed their genomes to their modern phase.

Other biologists have produced successful laboratory hybrid experiments among butterflies, frogs, fish, and marine invertebrates, and hybrids are being increasingly observed in nature, where success of the offspring usually results from crosses between closely related species. But the fusion of genomes from different phyla must surely result in an initial genomic chaos in the hybrid offspring, as hundreds, perhaps even thousands, of pre-evolved genes from two very distinct evolutionary pathways come together.

It seems likely, from what little we do know of the genomic responses to hybrid crossing, that complex internal mechanisms will have attempted to rationalise the potential outcomes. Key to any understanding of this complex situation are the developmental pathways that control not only form but also the differentiation and growth of internal structures, limbs, organs, tissues, and specific cells. Such developmental pathways will themselves be controlled not only by genetic factors but also by epigenetic controlling mechanisms – mechanisms involving the chemical control of the expression of genes – that are capable of responding rapidly to change in the genetic environment.[5]

Various outcomes are possible in such cross-phyletic crosses. Commonly, and most frequently seen in the crosses conducted by Holmes, Boerboom, and Williamson at Port Erin, there will be complete failure of development. Another possible outcome, as demonstrated in Raff's experiments among much more closely related hybrid crosses, is complex change in genetic regulatory pathways, with suppression of some pathways, enablement of others, and, most tantalizing, the creation of new pathways altogether. All of this would suggest that unknown genomic determinants and constraints might dominate at different times and in different situations, sometimes selecting pathways of maternal origin, sometimes pathways of paternal origin, and sometimes blending together the two disparate inheritances.

In science, however, we are obliged to confound any such interpretation in pursuit of an alternative. Could the results of the successful cross-phyletic experiments be a fluke, based on contamination with pre-fertilised echinoderm eggs? None of the three scientists who participated in the experiments view contamination as a likely possibility. Even if we discount their expert witness, accidental contamination would be expected to result in just the occasional success. Here the great majority of the eggs were observed to develop in a similar if wholly novel pattern, which makes the prospect of accidental contamination very improbable. Another rather unusual option, mentioned earlier in the Japanese studies, is a spurious development from the egg alone, a form of parthenogenesis known in the scientific jargon as gynogenesis. This would imply that the sperm, while triggering development, failed to merge its chromosomes with the maternal chromosomes in the egg. The subsequent offspring would be programmed entirely by the maternal genome, leading to a single-parent offspring. But an explanation based on gynogenesis would need to explain why, if development took place from the maternal genome, none of the plutei developed beyond the four-armed stage, why all resorbed their arms to become spheroids, and why some of these spheroids appeared to attempt a further development to the paternal ascidian body form. It would also need to explain how it was that many of the hybrid offspring entered a final phase of budding reproduction. There is no known biological or genetic explanation of how a development based wholly on the maternal sea urchin genome would give rise to a budding spheroid.

It would appear plausible that these experimental offspring were true cross-phyletic hybrids. This is the opinion of Don Williamson and the expressed conclusion of Sebastian Holmes. If so, the research offers yet one more astonishing potential implication.

The spheroids, in their capacity to reproduce through

budding, could be seen as a new, albeit artificially created, life form – arguably the first such lab-created life form in scientific history. This in turn presents several puzzles. Is it possible, for example, that the only viable offspring from such a disparate genomic union is an exceedingly simple one, perhaps created through massive suppression of developmental pathways? If so, is the spheroid a "rediscovered" animal form, the most basic of them – one that physically and reproductively resembles the nearest thing we might find to a common ancestor of the two distantly related phyla of marine invertebrates? Is it a clue to how multicellular life first came to inhabit Earth?

In the words of Scott F. Gilbert, 'one of evolution's most important experiments was the creation [from single-celled ancestors] of multicellular organisms'.[6] Ernst Haeckel proposed that the ancestor of the entire animal kingdom was a simple ball of cells surrounding a simple gut, the gastraea. But there is an obvious alternative: an even simpler organism, such as the simplest of all larval forms, the spheroid-like "planula", the larva of the jellyfish and medusa. The planula has no gut but imbibes nourishment through its surface layers.

For Francis Maitland Balfour, regarded with such high esteem by Darwin, the planula was not merely the earliest example of a primary larva, it also represented the primal common ancestor of the animal kingdom. Forms resembling fossilised blastulae have been discovered in Chinese and Siberian rocks; they may represent embryos of some of the very earliest animals and date to between five and six hundred million years ago.[7] Yet these may be eclipsed in age by much earlier spherical forms in fossils, dating to 1.2 billion years ago, found in the Stirling Range, in southwestern Australia, that may have been the oldest ancestors of multicellular life, such as animals and plants.[8] Other equally ancient fossil forms resembling embryos and planula larvae have been reported,[9] though these findings have been disputed.[10]

Williamson's extraordinary cross-phyletic hybridisation experiments deserve to be repeated by colleagues equipped with state-of-the-art molecular and genetic technology, so that the outcome can be assessed with the same intensity of scrutiny, at every stage of development, as seen in the work of Rudolph Raff and his colleagues. How stunning it would be if further experiment, backed up by epigenetic study of developmental sequences, were to confirm that Williamson's spheroids are not merely hybrids across different phyla, but might also be a physical recapitulation of our earliest multicellular ancestors from so very long ago.

# Part IV
# The Molecular Age

*After the structure of DNA was solved ... again and again, the smallest, most casual beginnings – a sceptical question asked in a Paris café, some specks seen on an electron micrograph, the idea struck off in a sentence during a drive down from New England – have grown up into specialties that have engrossed entire lifetimes in science, today command whole teams and laboratories, whose results fill volumes and are not ended.*

Horace Freeland Judson,
'The Eighth Day of Creation'

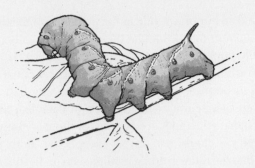

*Tobacco hornworm*

## 20

# The Puzzle of the Hornworm Brain

ON SEPTEMBER 30, 1999, the publication of a scientific paper in *Nature* caused a flurry of excitement, which spilled over into the US media. The title of the paper was "The Origins of Insect Metamorphosis", and the work of its authors, James W. Truman and Lynn M. Riddiford, based in the Department of Biology at the University of Washington, extended the earlier experiments conducted by Vincent B. Wigglesworth, Carroll Williams, and the many other experts involved in the search to understand metamorphosis.[1] The reason for the excitement: the paper questioned some of the most basic conclusions drawn by most scientists thus far.

We saw earlier how Wigglesworth's interest in metamorphosis lay not in its evolution but in its physiological control. While juvenile hormone and ecdysone provided a vital understanding of the physiological control mechanisms, and must surely have played an important role in the evolution of metamorphic change, attempts to extrapolate their function to other mysteries, notably pupation, remained unsatisfactory. Like Darwin, Wigglesworth had played down the differences between complete and incomplete metamorphosis, assuming that the dramatic, even cataclysmic, changes seen in complete metamorphosis – changes that appeared to mirror those of many marine invertebrates – should be seen as exceptional. Both scientists assumed that the more gradual change seen in the incompletely metamorphosing insects was more representative.

As Assistant Professor, Riddiford had co-authored a number of earlier papers with Carroll Williams, and both she and Truman had worked with Williams at his laboratory at Harvard.[2] But now they were exploring metamorphosis from a different angle than that taken by the earlier researchers. Where Wigglesworth, Willams, and their colleagues were largely interested in the physiology of metamorphosis, Truman and Riddiford were interested in a new avenue of biological thinking, part of a field that brought together the disciplines of evolutionary biology and development into the single discipline of evolutionary development. The youngest of the evolutionary disciplines, "evo-devo" was granted its own division in the Society for Integrative and Comparative Biology (SICB) in that same year, 1999.[3]

Evo-devo studies how development – whether embryological or post-embryological – affects evolution. This has obvious relevance to the evolution of metamorphosis, which is essentially post-embryonic development. In a recent review of how the field works, the American biologist Sean B. Carroll makes the point that two-thirds of the gene sequences exposed to natural selection in the human genome are non-coding – they don't translate into proteins that form part of the physical, or chemical, structure of the organism.[4] The existence of this vast, mostly unknown genetic territory might lead to unrealistic expectations about what can be learned from comparisons of protein-coding genome sequences alone. In fact a great many genes play a covert role in coding for regulatory elements – genetic sequences that direct embryological development, or, in adult life, the expression of protein-coding genes. The origins and functions of these regulatory sequences are of major evolutionary importance.

In their 1999 paper, Truman and Riddiford reappraised insect metamorphosis in a series of logical steps. 'Metamorphosis', they explained, 'is one of the most widely used life-history strategies of animals'. They agreed with the principle, espoused by the eminent Danish zoologist Claus Nielsen, that differences between larval and adult forms allow the same organism to exploit very different habitats and food sources.[5] In marine invertebrates, we have seen how metamorphosis allows a single organism to take advantage of planktic dispersal as a larva and then food resources and establishment of territories as a bottom-dwelling adult. In amphibians, it allows the organism to take advantage of marine and terrestrial habitats; in completely metamorphosing insects, it allows the adaptation of the plant-eating grub or caterpillar larva to the winged adult, which is often non-feeding, the latter providing sexual reproduction as well as assisting dispersal to the widest possible geographical range. Truman and Riddiford also disagreed with Williamson in sharing the view of Nielsen and another Scandinavian, Gösta Jägersten, that metamorphosis in amphibians and many marine invertebrates was an integral part of the animals' earliest life cycle.[6]

But this did not apply to insects. We know that the first insects, which originated in the Silurian period, some 438 to 408 million years ago, were wingless and did not undergo metamorphosis. This means metamorphosis must have been introduced into the life cycle through some later evolutionary adaptation. As Truman and Riddiford now explained, insects with incomplete metamorphosis – for instance, cockroaches and dragonflies – had mixed evolutionary origins. In other words, incomplete metamorphosis evolved again and again, such multiple origins, in the scientific jargon, being "polyphyletic". We might also recall, from the example of *Rhodnius prolixus*, that nymphs, the larval equivalents among incompletely metamorphosing insects, look like immature adults. Besides lacking genitalia, nymphs have wing buds folded over their backs, and these wing buds transform into the fully functional wings during the final moult to the adult.

Truman and Riddiford also highlighted some essential differences in the evolution of incompletely and fully metamorphosing insects. 'Insects with "complete metamorphosis" were first seen in the Permian', from 286 million to 245 million years ago, they wrote, 'and constitute a monophyletic group' – that is, a group with a single evolutionary origin. This implied that a single evolutionary event gave rise to the wonder of butterflies, beetles, moths, flies, and bees.[7] This variance in the origins of incompletely and completely metamorphosing insects brought into question a central assumption that Wigglesworth had made at the beginning of his research – an assumption by and large accepted in entomology. He had assumed his findings about the incompletely metamorphosing bug *Rhodnius* would be equally relevant to fully metamorphosing insects. This physiological perspective had proven useful to Wigglesworth, enabling him to simplify his experimental designs. But from an evolutionary perspective, the approach did not fully explain the process of complete metamorphosis, during which catastrophic changes occur within the pupa. How, Truman and Riddiford wondered, did complete metamorphosis evolve from the simpler life cycles of the ancestral insects? This demanded further investigation.

Perhaps more astonishing, in questioning the prevailing viewpoint they reopened a controversy about metamorphosis that had begun some 2300 years earlier with Aristotle, who had put forward the theory that the caterpillar was nothing more than a soft egg.

It all began with a romantic story. Lynn Riddiford first met her future husband, James Truman, at Harvard University in 1967, soon after Truman joined Carroll Williams's department as a graduate student. In Truman's words, 'As it turned out, I was assigned Lynn as my graduate adviser. So I got my degree working with her. After I got my degree, we were married and I stayed on for an additional three years in the department.' That is a modest way of summing up a chance meeting that would give rise to one of those creative partnerships that have been described as the marvellous resonance between two very original minds.[8]

Riddiford did not begin her scientific career with an interest in metamorphosis. Originally her attention ranged over a wide variety of subjects, from physical chemistry to the ability of newts to regenerate lost limbs. But she switched to metamorphosis when she was an undergraduate working with Williams. 'He'd just given us a talk about juvenile hormone, which I had never heard of before. I asked him the odd question or two. For example, I wanted to know if juvenile hormone had anything to do with the metamorphosis of tadpoles. He said he didn't know – but would I like to try to find out? I investigated it and found that it didn't have any effect. But anyway, that's where my interest began.'[9] Truman, on the other hand, had always been captivated by the process. Inspired by Fabre, he had become an avid insect collector while still in his early teens and devoured the French naturalist's books. In particular, he loved *The Hunting Wasps*. As an undergraduate at the University of Notre Dame, in Indiana, he became involved with mosquito research, before moving on to his graduate education under Williams and Riddiford.

I asked him, 'What was it about insects that so interested you?'

'It was the diversity. I was a bit of a morphology freak. In one sense, insects have a very simple body plan – but, thanks to this amazing cuticle, they can make such a diversity of shapes and patterns. I went through a butterfly phase. But over time butterflies just didn't excite me that much any more. Morphologically, they're quite similar, although the colours are beautiful, quite fantastic. I was far more interested in beetles, because of the huge morphological diversity.' Darwin had arrived at the same conclusion. Amazingly, there are about half a million species of beetles with myriad varieties of shapes, colours, and living strategies.

While working for his PhD in Riddiford's lab, the youthful Truman made his first discovery. The hormonal control mechanisms of metamorphosis were already established by then, but there was no hint that the behaviour of insects at the time of moulting might also be controlled by a hormone. Truman's earliest work, under the tutelage of Riddiford, involved the giant silk moths favoured by Williams, but in time he switched his experimental subject to the sphinx moth, *Manduca sexta*, whose larva is the tobacco hornworm. One of the familiar garden pests of America, the hornworm is pea-green in colour with white stripes running along its sides; growing to a gigantic ten centimetres in length, it can quickly defoliate crops, including tobacco, tomatoes, potatoes, aubergine, and peppers. It is also recognisable from the prominent red spike at its rear, curved like a horn, that lends it its name. When Truman adopted this giant caterpillar as the subject for a series of investigations of moulting behaviour, he discovered that a brain-sourced hormone – the "eclosion" hormone – was responsible for larval behaviour at the time of moulting. Over subsequent decades Truman and Riddiford went on to locate the cells in the insect brain that secrete the hormone, its amino acid sequences, and finally the gene that codes for its sixty-two-amino acid peptide sequence.[10]

Today we know that the eclosion hormone is released during a precise stage of the moulting cycle, when it acts directly on the brain to trigger moulting behaviour, leading in turn to tissue changes and the behaviour that enables the shedding of the old cuticle.

The eclosion hormone appears to be unique to insects, and the diversity of its effects are still being evaluated. Finding the eclosion hormone was just one of many important discoveries made by these two remarkable biologists. The range of their investigations covers many aspects of insect development, metamorphosis, and evolution. In one investigation, the two scientists looked at the hormonal control of migratory flight in the large milkweed bug.[11] Studies such as this are now helping us understand the annual monarch butterfly migration, which, in Truman's opinion, is most likely cued by the shortening period of daylight, and probably linked to changes in juvenile hormone secretion.[12]

In the study of development, most of their fellow entomologists had been content to accept that the mystery of insect metamorphosis had been solved by the discovery of the controlling hormones. But Truman and Riddiford, who had worked in the laboratory of one of the pioneers, felt differently. In their subsequent careers, they have accumulated a vast knowledge and experience of insect development, and, by implication, insect evolution. In the course of their research, however, they encountered numerous instances of developmental phenomena that appeared to make no sense according to the prevailing theories. In Truman's words, 'This became glaringly evident from the nervous system'.

In the mid 1980s, Ron Booker, a post-doctoral researcher working under Truman's guidance, was studying the development of the nervous system in the caterpillar of the sphinx moth. In the nerve clusters which comprise the thoracic ganglia, Booker came across stem cells that generated large numbers of new neurons as the larva grew, but these neurons remained dormant in the caterpillar, only coming to maturity during the metamorphosis to the adult moth, when they produce most of the adult nervous system. Stem cells that develop into nerve cells are known as neuroblasts. A little later, when Truman and Riddiford took a sabbatical to work in the lab of Mike Bate at the University of Cambridge, Truman found similar stem cells in the larva of the fruit fly. This confirmed that such neuroblasts were crucial in the metamorphosis of the moth and the fly. Truman wanted to know more about them. Where did they come from? 'Were they a new set of stem cells unique to larvae that undergo complete metamorphosis? Or were they stem cells that had persisted from the embryo stage – in other words, were they embryonic stem cells that had lain dormant in the embryo only to activate in the larva to make the rest of their lineage of neurons?'

Truman appreciated that in incompletely metamorphosing insects, such as grasshoppers, similar stem cells generate all of the neurons in the embryo prior to hatching. After this, they appear to die off, so that, by the time these insect varieties reach the nymphal stage, their ganglia contain no residual neuroblast stem cells. This suggested a major difference in the metamorphic development between completely and incompletely metamorphosing insects.

After their time at Cambridge, Truman and Riddiford travelled to Nairobi, where they turned their attention to tsetse flies, the insects that carry the trypanosome parasite that causes African sleeping sickness. Tsetse flies, like sphynx moths and fruit flies, undergo complete metamorphosis. Again the partners found those same stem cell neuroblasts, which they could track from their appearance in the early embryo and into the larva, when they switched to making the neurons for the adult version of the nervous system. (Other researchers later confirmed these discoveries through elegant transplant experiments performed on fruit flies.[13]) By now Truman and Riddiford were convinced that they were seeing a pattern, and that this pattern pointed to an important variance between complete and incomplete metamorphosis in insects.

This conviction was at the back of their minds when, in 1996, they flew to Canberra to spend several months working with Eldon Ball at the Australian National University.

*Pronymph of milkweed bug*

## 21

# Aristotle or Darwin?

IT WAS ACTUALLY THE SECOND trip to Australia taken by Riddiford and Truman. Three years earlier they had worked, respectively, at the Commonwealth Scientific and Industrial Research Organisation (CSIRO) and the Australian National University, during which time Riddiford focused on the fruit fly and Truman on insect embryos. This earlier experience had given them a better appreciation of the embryonic development of simpler insects, such as grasshoppers.

This new trip came as a breather from their normal routines, allowing them time and opportunity to think about where all of their work was heading. In spite of the anomalies, they still accepted the prevailing views of earlier colleagues, and of Wigglesworth and Williams in particular. But they also knew that, somehow, the wonder of metamorphosis had interposed itself into the life cycle of those first insect species that had evolved to colonise dry land without the need or advantage of metamorphosis. The partners were also increasingly questioning the generalised mode of scientific thinking that lumped together completely and incompletely metamorphosing insects into the same evolutionary, and developmental, explanation. Even today four of the most primitive orders of insects are wingless and still do not metamorphose, including the springtails, which include the earliest fossil insects. Now, in Truman's words, they arrived at the decision: 'Maybe we should look at the embryonic development of the more basal insects.'

They turned their attentions to two familiar examples, silverfish and locusts. Silverfish belong to the order Thysanura, also known as bristletails, a primitive group of white, eyeless, and wingless insects that inhabit soil, rotting wood, and the crevices of kitchens and bakeries. These insects retain rudimentary appendages on their abdominal sections, which may be a remnant of a still more primitive form with limbs across all of the body segments, like those seen in millipedes and centipedes. Silverfish grow through moulting, but do not undergo metamorphosis. Locusts are members of the order Orthoptera, which includes grasshoppers and crickets, with fossil members that date back to the Carboniferous period, some 360 to 286 million years ago, and which undergo a basic form of incomplete metamorphosis.

Truman and Riddiford wrapped batches of eggs of these two primitive insect species in small pieces of cheesecloth and dropped juvenile hormone, dissolved in a little acetone, on top of them. They didn't need to perforate the egg membrane because the hormone readily diffused through the membrane and entered the embryo. Some of these eggs were transparent, so they could transilluminate them and photograph what was happening inside the membrane. The results were startling: 'Suddenly', in the words of Truman, 'we began to see that a lot of the issues we had been concerned about in metamorphosis came into play in the embryos of these more basal insects. That made us start to think afresh about the current notions people were holding.'

The non-metamorphosing silverfish and incompletely metamorphosing locusts hatch out from the eggs as nymphs, which closely resemble fully-grown adults. Wigglesworth, who more than any other scientist had established the paradigm of insect development, had assumed that the larvae of fully metamorphosing insects were equivalent to the nymphs of non-metamorphosing and incompletely metamorphosing ones. The prevailing modern hypothesis further assumed that, as the disparity between the larval and adult stages of fully metamorphosing insects grew – through the action of natural selection on two very different stages and life cycles – there was a biological need to "bridge" the widening gap in order for the adult to emerge. This bridge, over evolutionary time, became the pupa. A purist might baulk at the concept of "need", since evolution is a chance affair paying no heed to such a human concept. An evolutionary biologist might better express this as a selection pressure for a transitional form. But this was a hypothesis, however expressed, that Riddiford and Truman now felt compelled to challenge. An alternative explanation that might better fit the evidence lay hidden, largely forgotten, in the murky pages of history.

When Aristotle argued that the caterpillar was a soft egg, he meant that it was a continuation of embryonic development. In 1651, at Wigglesworth's cherished Caius College, William Harvey took the ancient philosopher's thinking further, supporting his idea of the caterpillar as a free-living embryo, but suggesting that the *pupa*, and not the caterpillar, of the butterfly was the equivalent of the nymph of an incompletely metamorphosing insect. In 1913 the Italian naturalist Antonio Berlese extended this still further, proposing that metamorphosis arose through the hatching of the insect larva at a premature stage of embryonic development – a view referred to as the "de-embryoniza-tion theory".[1] In the first half of the twentieth century, Berlese's approach was developed by J.J. Jeschikov, who pointed out that there was a close link between the amount of yolk in the insect egg and the timing of hatching.[2] If there was less yolk, say in the butterfly egg, the larva hatched out early as a caterpillar. For a time, the de-embryonization theory, also known as the Berlese-Jeschikov theory, captured the imagination of biologists.

Wigglesworth disagreed with these and other alternative theories.[3] Darwin had earlier proposed that metamorphosis was the development of a single life form embracing a number of different body forms, or polymorphisms, so that each body form was a step, or plateau, along the developmental pathway. This allowed natural selection to adapt each stage along lines that fitted the selective pressures of that stage. Wigglesworth agreed with this Darwinian explanation. 'The most likely sup-position is that there has been an independent evolution of the different "stages" of insects', he said.[4] This organism had some-how evolved the ability to change its form at different periods of its development. He believed insects contained in their genetic make-up the potential for these different body forms. This does not imply that, in the language of science fiction, they were genetically controlled to be shape-shifters. Rather it presaged the modern evo-devo interpretation: the distinct body forms were determined by the organism's genetic make-up, operat-ing through a series of discretely inherited developmental path-ways. Given such a perspective, we can see why Wigglesworth saw no essential difference between the process and its evolu-tion in all metamorphosing insects. The incompletely meta-morphosing insects had evolved the potential for nymph and adult forms, whereas the completely metamorphosing insects had, through very similar mechanisms, evolved the potential for larva, pupa, and adult.

If both Wigglesworth and Darwin were right, this is as good

an explanation of the evolution of metamorphosis we're going to get. The evolution to complete metamorphosis, a more gradual process than occurs in incomplete metamorphosis, provided a means for wiping the slate clean so that, within the pupa, it could start afresh from "set-aside" stem cells and effect a phoenix-like rebirth.

To Truman this hypothesis seemed too conceptually nebulous – and moreover it failed to explain several glaring anomalies in light of his experience. So he considered some alternative proposals, including those of Harvey, Berlese, and other successors to Aristotle. 'The essence of the Berlese theory – as with Harvey and Aristotle – is that the larva is a feeding embryo, or [to paraphrase Aristotle], a feeding egg', Truman explains. 'That's very nice but the real issue is what really happens if an embryo hatched prematurely in this way? The same thing would apply to a baby born prematurely. The maturation isn't finished and the animal would not be viable. Therefore, if you are going to [propose a theory in which] an animal hatches at an earlier embryonic stage, [you're going to have to explain how it has come to] be developmentally functional.' Together, Truman and Riddiford thought it worthwhile to probe the evo-devo conundrum.

In metamorphosis juvenile hormone is a "status quo" hormone, a term coined by Truman's former professor, Carroll Williams. It blocks development, keeping the insect in the same form as it was prior to a moult. Truman and Riddiford found that in the silverfish and locusts there appeared to be an altogether different role for the same hormone while the embryo was still forming within the egg. 'When we added juvenile hormone during embryonic development, we found that instead of keeping the embryo in status quo – as an embryo – it actually caused the embryo to undergo premature maturation'. This was a startling revelation.

When the two scientists went back and hit the journals in the university library, they found that it should not have come as such an original surprise. There was already some information in the scientific literature to suggest that the embryonic effects of juvenile hormone varied between basal and more advanced insects. People had simply ignored these earlier findings, leaving it for Truman and Riddiford, in a sense, to rediscover the wheel. But as Truman expressed it to me, 'It also involved our looking at the wheel in a different way.' They realised this discovery potentially had evolutionary significance.

When they treated silverfish eggs with juvenile hormone, it had a devastating effect on embryonic development. In the insect embryo, the limbs begin to grow out from appropriate segments along the body axis. Normally they begin as extensions from the body and then they segment, to give rise to the typical multi-jointed insectile legs. This is followed by the laying down of the mature skin layer, or cuticle, and by the development of the necessary muscles of attachment. When Truman and Riddiford added juvenile hormone to developing embryos of silverfish – say, right after they first developed limb buds – the embryos laid down the mature cuticle on the buds, halting further development of the legs. The exposure to juvenile hormone caused a premature maturation, which resulted in an embryo that was not viable. This implied that during the normal embryology of these basal insects, juvenile hormone would not normally come into play until the very end of development. When, on the other hand, they repeated the experiment with more advanced insects, such as incompletely metamorphosing grasshoppers and crickets, they discovered that juvenile hormone had no such effect at this early stage in development. The effects of juvenile hormone only kicked in when they reached a stage close to hatching. Then the hormone caused a premature hatching, but in these more advanced orders it took place at a mature enough stage to give rise to viable animals.

In mammalian pregnancy, birth is very precisely timed, as

is the hatching from the egg of birds, reptiles, and insects. The developmental timing of hatching is critical, as the newborn or hatchling emerges from the protected environment of the womb or egg into the less protected environment of the outside world. Nevertheless, across the spectrum of animals, birth or hatching takes place at very different stages of development, including in some what is clearly an intermediate developmental stage.[5] For example, the kangaroo young emerge from the womb while they are still the equivalent of embryos and complete their development in the maternal pouch. In many spider groups, the hatchling larva is typically non-feeding, lacks pigmentation, has limited locomotion, and continues to feed off its yolk stores; in scorpions, the non-feeding hatchlings are carried for up to two weeks on the backs of the mothers. Nowhere is the hatching of offspring at embryonic stages more obvious and startling than the larval forms of the many aquatic animal phyla, with sea urchins, for instance, emerging as living blastulae. Perhaps it is not surprising that a specialised hatchling stage might also be seen in the more basal orders of insects.

There is no general name given to this premature hatchling stage in insects, so Truman and Riddiford decided they would adopt the name used in the dragonfly literature – the pronymph.[6] In the life cycles of the non-metamorphosing and incompletely metamorphosing insects, as they now proposed, there exists a largely ignored stage, this pronymph, which occupies a period between the development of the embryo in the egg and the emergence of the adult or, in the incompletely metamorphosing insects, of the nymph. The pronymph has a number of characteristics that make it unique – the proportions of its body, the microscopic structure of its cuticle, its incompletely developed mandibles, and so on; in other words, it is significantly different, and simpler in structure, than the nymph. The scientific partners now proposed that the pronymph of these more primitive orders of insects, including the basal orders and some of the incompletely metamorphosing orders, such as the dragonflies, was the evolutionary ancestor of the larva of insects that underwent full metamorphosis. They also proposed that the nymphal stage of the incompletely metamorphosing insects had evolved into the pupal stage of completely metamorphosing insects.

The entomologists admitted that there were residual problems with their proposal. The pronymph lasts for only a single phase in the life cycle of the basal insects and dragonflies, while larvae progress through a succession of moults and stages. But modifications of hormone control might explain these differences. Their theory, which amounts to a renaissance of the idea first put forward by Aristotle, could be seen to fit with what is already known to happen in both incomplete and complete metamorphosis. In incompletely metamorphosing insects, such as dragonflies and *Rhodnius prolixus*, the first withdrawal of juvenile hormone results in metamorphosis to the adult. In completely metamorphosing insects the situation is more complex, with a small and then a larger ecdysone peak, the first priming certain tissues to get ready to change to those of the pupa, and the second giving rise to the pupal moult. Juvenile hormone is absent at the beginning of the pupal moult; it subsequently reappears during early pupation only to disappear again before the ecdysone peak that gives rise to the formation of the adult insect within the pupa.

In 2004, this hypothesis received its first molecular confirmation, when Deniz Erezyilmaz, a doctoral student working under their direction, examined the effects during development of a gene called the broad complex.[7] Riddiford's lab had already shown that this gene controlled the expression of the pupal stage. For example, during a larval moult in the fruit fly, the larval tissues begin to activate pupal genes. When the broad complex is expressed in what should be the final moult to the adult, instead of activating adult genes the tissues revert to activating pupal genes. So the broad complex appears to specify development of the insect into a pupa. It is not activated – indeed, it could not be activated – in the larval-to-larval moults of completely metamorphosing insects, such as the grub of a bee or the caterpillar of a butterfly, because it would turn the larva into a pupa. When, for the first time, Erezyilmaz examined the broad complex's role in the more basal insects, she came across a revealing characteristic: whereas in the completely metamorphosing insects the broad complex was only expressed in the pupal stages, in the incompletely metamorphosing insects the same complex was expressed in the nymphal stages. This supported the Truman-Riddiford theory, which equated pronymph to larva and nymph to pupa.

*Red abalone*

## 22

# Cues and Common Links

ON JANUARY 26, 1971, while researching his book *The Eighth Day of Creation*, the acclaimed historian Horace Freeland Judson travelled to the University of Cambridge to interview the molecular scientist Sydney Brenner. Over dinner that evening at King's College, the conversation turned to the ways in which science was changing. Brenner, one of the elect few at the centre of the DNA story, commented that biologists can only be interested in three basic questions: 'How do things work? How are they built? How do they evolve?'[1] Much as Wigglesworth saw physiology at the root of all questions, to Brenner's mind it was the second of the three questions that took precedence. It was '… the deeper question that comes before you can answer the evolutionary question: How are organisms built?'

One of the wonders of nature, and closely related to the mysteries of metamorphosis, is how a single fertilised egg develops into an oak tree, a blue whale, or a human baby. Professor Eric H. Davidson at the California Institute of Technology has played a major role in investigating the genetic control of development, and in particular has taken a directorial role in unravelling the sea urchin genome.[2] In his book *Genomic Regulatory Systems*, Davidson acknowledges the importance of Darwinian evolution before stating that 'classical Darwinian evolution could not have provided an explanation, in a mechanistically relevant way, of how the diverse forms of animal life actually arose during evolution, because it matured before molecular biology provided explanations of the developmental process'.[3] It is important to grasp that Davidson is literally writing about 'forms', that is, the shapes and physical structures of living organisms. And by 'Darwinian evolution', he is referring not to Darwin's theory itself but to the synthesis theory of the 1930s, in which, he notes:

> ... *the argument was that organismal evolution is the product of minute changes in genes and gene products, which occur as point mutations and which accumulate little by little, providing the opportunity for selection and ultimately reproductive isolation. The major forms this argument has taken have focused on stepwise adaptive changes in protein sequence, but this is probably largely irrelevant to the evolution of any salient features of animal morphology.*

Davidson proposes that changes in the DNA sequences that control developmental processes, as opposed to changes in the DNA of genes that code for proteins, have played a fundamental role in the evolution of form. Sean B. Carroll makes the same point when he explains, 'regulatory sequences are so often the basis for the evolution of form that, when considering the evolution of anatomy (including neural circuitry), regulatory sequence evolution should be the primary hypothesis considered'.[4]

The work by Truman and Riddiford on the effects of hormones on insect embryos is a telling example of this line of thinking. Understanding the evolution of metamorphosis, including its hormonal triggers and the resulting genetic cascades, affords insight into how organisms are built – one of Brenner's top questions. How such developmental pathways work, how their biochemistry and physiology are rigged, and how they evolved, are the very essence of the story of evolution in general, and the evolution of metamorphosis in particular. To learn more, we could do no better than return to the pioneering Vincent Wigglesworth and his study of the blood-sucking insect, *Rhodnius prolixus*.

Adult females of this species, as we've seen, produce new batches of eggs each time they drink blood. It's the blood meal that acts as the primal signal, or cue, for each successive moult. There is a physiological explanation for this. Blood proteins supply the amino acids needed by the female for egg production, in particular for the manufacture of the yolk protein, also called vitellogenin.[5] As Wigglesworth demonstrated, it is the physical stretching of the larval abdomen that signals the brain to activate the hormonal triggers for metamorphosis. Juvenile hormone also stimulates yolk protein synthesis in the insect ovary.[6] And what works for Rhodnius also works, with perhaps some variation, for other insects. For example, in many mosquitoes, egg production is also triggered by a blood meal. Only female mosquitoes feed on blood. Without the blood meal, they do not make yolk protein. In *Aedes aegypti*, the insect that spreads the yellow fever virus, the digested products of the blood meal stimulate the insect brain to secrete egg development hormone, which in turn stimulates the ovary to make eggs.

Environmental cues are also important in marine metamorphoses. These triggers are numerous in type and can vary between species within the same genus or family. Detection of the cue, which depends on a specific sensory system, ensures that the larva settles in a habitat suitable for survival and growth – often a firm surface and a convenient food source, such as a specific alga. While some signals encourage settlement and metamorphosis, others inhibit it, so that the larva must take into account a complex cocktail of conflicting signals within a confined environment.[7]

Take the Californian red abalone, *Haliotis rufescens*. The largest abalone in the world, and historically the most important commercial species on the west coast of the United States, this impressive mollusc inhabits intertidal and subtidal rocky areas along the Pacific coastline, where the adults use their powerful feet to cling to rocks, allowing them to graze on kelp and other algae. The adults reach sizes upwards of thirty centimetres and can live for as long as fifty-four years. The great size and the succulence of the abalone's flesh led to such overfishing that the species is now protected from commercial exploitation, although recreational fishing is still allowed off the northern California coast. The mollusc's striking red colour derives from the same pigment that is found in red algae. Indeed, so beautiful is the shell that Native Americans used its colour, spiral form and mother-of-pearl interior to construct the pendants and ornaments used in shamanistic Kuksu dances.

Biologists have taken a special interest in the red abalone's natural history and life cycle. The mollusc reproduces through separate sexes and its eggs are fertilised externally. Mature females lay up to fifteen million eggs annually, which hatch into free-swimming trochophore larvae, about two millimetres in diameter. The trochophore metamorphoses to a more complex "veliger" larva, which, though only slightly bigger, develops two ciliated flaps that enable locomotion, and thus reduce the risk of being eaten. The veliger increases in size, develops more complex internal organs, secretes a protective shell, and grows what will become its future, often iridescent, foot. During this series of changes, the larval body undergoes a twist through

180 degrees on its longitudinal axis. After another two to three weeks, it metamorphoses to the juvenile adult, which abandons its planktic existence and settles to the ocean floor.

We might pause to reflect on the fact that the abalone has four distinct periods of development: first as the embryo within the egg, then as each of the two larval phases in turn, and finally as the adult. Each development requires different genetic programming. These must have evolved separately, whether through natural selection operating on distinct phases of development within the same organism, or, if Donald Williamson is right, through selection operating on entirely different organisms that subsequently hybridised to the single genome of what we now recognise as the abalone – or possibly a combination of both. The veliger of the Californian red abalone also settles in response to a specific environmental cue, which kicks in when it first makes physical contact with coralline red algae. The briefest of contact is all it takes. It would appear that a receptor on the larva recognises a highly specific chemical on the surface of the alga. So vital is this environmental cue that contact with the alga triggers metamorphosis in the larva, which stops swimming and begins the final development to the juvenile adult.

Another interesting example of environmental triggering is the symbiotic incorporation of bacteria in the bodies of insects and animals. The luminous bacteria *Vibrio fischeri* is essential to the development of the juvenile Hawaiian bobtail squid, *Eupyrmna scolopes*; the squid cannot develop without them.[8] Another example comes from the world of insects. In the leafhopper, *Euscelis incisus*, symbiotic bacteria within the eggs are transferred to future generations. If the bacteria inside the eggs are killed off, the embryos fail to form a gut. The environmental cues that trigger and control this development are not necessarily chemical. Sometimes they are physical, for example the influence of gravity, or the effects of ambient temperature, light duration, and humidity, which change with the seasons.

How does the internal machinery of life respond to these environmental signals? Very likely this involves a range of different molecular and genetic mechanisms. We know that the environmental cue initiates some key process of internal change. It has come as something of a surprise, and a wonderful surprise at that, that in plants one such response mechanism involves non-DNA-based epigenetics. This controls the activation and expression of genes, allowing environmental signals to trigger genes in response to, say, the arrival of spring.[9] In insects, of course, this takes the form of a hormonal response that in turn initiates the genetic cascades that programme development.

Even before the discovery of the structure of DNA, scientists had observed a curious phenomenon that took place on the chromosomes of fruit fly larvae. Imagine the chromosomes as incredibly long strings of beads, each individual bead exceedingly narrow and with hundreds, sometimes even thousands, of beads being strung tightly together along a single chromosome. Today we know that these beads, which in insects are visible with a light microscope, represent genes. During metamorphosis, specific beads at some points along the string of the chromosomes begin to swell up, so they resemble woolly scarves, an event known as chromosome puffing. As long ago as 1952, the German biologist Wolfgang Beermann suggested these chromosome puffs were the site of developmental gene activity.[10] The puffs can be so dramatic they increase the bead width four-fold; in the process, they accumulate high levels of RNA and protein, indicating that the puffed-up genes are being expressed. Subsequently, new techniques in molecular chemistry have confirmed the puffs do represent genes being expressed as messenger RNA, the molecule that carries the nuclear coding of DNA from the nucleus to the protein-manufacturing factories of the cell. The DNA of genes is condensed within chromosomes by packaging it around small circular proteins, called histones. For a gene to become active, its binding to the histones needs to be loosened. Puffing is the visible manifestation of this.

In the fruit fly, towards the end of the third larval stage and just prior to pupation, a new set of six chromosome puffs appears. These are dubbed the "early puffs", and they regress after three or four hours. They are followed by a set of over a hundred "late puffs", which also regress in a very specific time sequence. The cycles of puff appearance and regression on various chromosomes are so regular that scientists are able to correlate their situation and timing with specific aspects of metamorphosis, for instance with the synthesis of an important protein or with the timing of the eversion of the head from the thorax, which happens during the transformation to an adult. The observation of puff cycles offered the first clue that development – during metamorphosis or during embryonic development within the egg – depends on the activation of specific genes in a tightly controlled sequence.

In the early 1960s, biologists showed that they could induce typical puffing of chromosomes by injecting purified ecdysone into the blood stream of insect larvae at a stage prior to metamorphosis to the pupae.[11] The biologist Ulrich Clever observed, when he and his colleagues injected ecdysone into the fourth stage larva of the water midge, *Chironomus tentans*, that two puffs were formed within thirty to sixty minutes, another two appeared at five to twenty hours, and two more at forty-eight to seventy-two hours. The location and timing of these puffs was precise and reproducible. This predictability was later confirmed in fruit flies and other insects. It was the first clear evidence that the hormones discovered by Wigglesworth, Williams, and others, in particular the moulting hormone, ecdysone, work by initiating a precise genetic pattern that leads to sequential cascades of gene activation during metamorphosis.[12] Now that ecdysones have been identified as steroid hormones, in other words, as members of a group of endocrine regulators that are found throughout all of life, we can begin to see how it might affect development beyond the metamorphosis of insects.

The active form of ecdysone, known as 20-hydroxyecdysone, cannot attach to the DNA that makes up a gene. Activated ecdysone must bind to a site on the chromosome close to the gene on which it is to be targeted. There are a number of control sites, called receptors, located conveniently near to important developmental genes, sites that can activate, repress, or otherwise influence the expression of the gene. When ecdysone arrives at the ecdysone receptor in insect chromosomes, metamorphosis is triggered.[13] In some cases, this causes the early puffs to regress – a repressive action; in others, it induces rapid puffing – gene activation. Late puffs are often resistant to ecdysone, indicating they are preferentially responsive to the cascade of new developmental activators and repressors "downstream" of the initial action of the hormone. Many, though not all, of the genes stimulated by ecdysone trigger a complex cascade of responses, with activation in some genes and repression in others. This precisely-timed programme prompts the destruction of certain tissue and organs specific to the larva and the formation of the tissues and organs specific to the adult.

How fascinating, then, that the ecdysone receptors in the chromosomes of insects are almost identical in structure to the thyroid hormone receptors in chordates and vertebrates! This provides an important clue to the mysteries of vertebrate evolution and development. Over recent decades, molecular biology has extended our knowledge of these developmental pathways, so often initiated by hormones, to show how, at the level of genetic action, the downstream ramifications are formidably complex. It seems likely that similar triggers, systems of response, and subsequent developmental mechanisms control metamorphosis in marine invertebrates. It appears that Wigglesworth was right when, at the very beginning of his career, he suggested that understanding insect development and physiology would assist our understanding of the development and physiology of all life.

Although there are more than three quarters of a million species of insects – and some believe many more – they constitute a single class within the phylum Mandibulata, the arthropods equipped with jaws. Perhaps it is not surprising that metamorphosis in insects has commonalities throughout the entire class. Marine invertebrates, by contrast, are far more diverse. Their evolutionary range extends to many different phyla and often they display body forms dating back to the Cambrian explosion, when all but a single animal phylum first appeared in the fossil record. This may explain why there is more diversity in the evolution of metamorphic controls and developments in marine invertebrates than there is among insects. In marine invertebrates, the embryonic phase can be very short and the larva often emerges at a very early stage of development – for example in the sea urchins, where the egg hatches as a blastula. This blastula then develops to a very simple larva: the wheel-like trochophore of annelids and molluscs and the easel-shaped pluteus of sea urchins. In marine groups, the adult is usually much longer living than the larva, with the larval phases adapted to dispersal, whereas in many completely metamorphosing insects, such as moths and butterflies, the very opposite applies, with an extended larval phase, devoted to rapid feeding and growth, and the adult phase, often brief and sometimes entirely non-feeding, devoted largely to reproduction. All of this would lead us to anticipate fundamental differences between metamorphosis in insects and marine invertebrates.

Insects do have a distant common origin with the marine arthropods, however. And one of the surprising and intriguing revelations of developmental biology is that developmental pathways have been remarkably conserved over vast stretches of evolutionary time. This might result in commonalities in the evolution of insects and related marine arthropods – and, perhaps even more widely, all marine groups. One obvious place to look is to the catastrophic metamorphoses found in both insects and marine animals. Are there similarities between the "set-aside" imaginal discs of the completely metamorphosing insects and the pluripotent stem cells that line the coelomic sacs of marine invertebrates? The answer appears to be yes – and strikingly so.

The next question, therefore, is how these pluripotent cells, programmed for very different developmental patterns from the larva, have evolved. Did they arise through mutations? If so, this would imply that mutation occurred at the very earliest stages of embryogenesis, before the dividing cells acquired any degree of differentiation. Or did they arise, even in part, through hybridisation, as Don Williamson argues? Both explanations involve a certain leap of faith, since they require that the genome has evolved ways of creating separate developmental blueprints within the same fertilised egg, and of expressing these separate developmental pathways in strict sequence during the metamorphic process, including the extreme examples, such as the sea squirt, where we witness the two blueprints developing simultaneously. And whichever mechanism one favours, whether mutation, hybridisation, or perhaps an amalgamation of several mechanisms, this would still require the evolution of some means for setting aside the pluripotent cells, so that they remain impervious to the developmental commands throughout the larval stages and only activate when it comes time for the animal to develop into an adult. The origins of set-aside cells are an ongoing conundrum for evo-devo experts, and its answer may involve epigenetic contributions, the heritable changes in genetic activation and expression that do not alter the DNA itself.

Minor modifications in developmental pathways can have major effects. But once a core developmental pathway has evolved, its very usefulness would appear to confer a resistance to change. This has been elegantly displayed in studies of Hox gene clusters, which play a major role in developments along the longitudinal axis of bilaterally symmetrical animals, and which show amazing conservation over vast periods of evolution. Moreover, over the long term, some degree of parsimony in the evolution of hormonal signalling also seems likely. In the preface to their recent collection of articles on the subject, Lawrence I. Gilbert and his colleagues highlight the fact that '… remarkably, the phenomenon of metamorphosis continues to be an excellent model for those interested in the control of development and endocrinological processes'.[14] And although amphibian metamorphosis looks very different to insect metamorphosis, here too we find evolutionary conservation of several mechanisms of hormonal control of development. For instance, the receptors for ecdysones and thyroid hormones are so similar that artificially reproduced chimerical forms of ecdysone hormones are effective not only in activating target genes in the cells of invertebrates, but also in vertebrates.

Our own vertebrate ancestors are presumed to have evolved through an amphibian stage. This would suggest we have something to learn about our own evolutionary development from looking at what happens among the amphibians, many of which undergo striking metamorphosis.

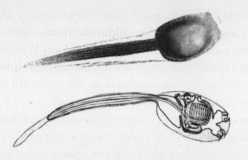

*Ascidian tadpole larva*

## 23

# A Tale in a Tail

VERTEBRATE EVOLUTION, with its chordate ancestry, has a surprisingly ancient history, and some biologists believe that the sea squirts, with their tadpole larvae, may represent the earliest known stage of that history. As we have seen, the sea squirts, which belong to the phylum variously known as ascidians or urochordates, are essentially headless creatures with the rudiment of a spinal cord. Frogs and toads, by contrast, belong to the phylum of the craniates, the more familiar chordates that do have heads. And while sea squirt larvae bear a remarkable resemblance to the tadpoles of frogs and toads, in particular in sharing the chordate locomotory tail, it would be disingenuous to equate the two. The sea squirt larva is only two or three millimetres long, is incapable of feeding, and possesses an internal structure that is much simpler than that of the amphibian tadpole. Moreover, its evolution vastly predates that of the amphibians. In China, giant tadpoles have been identified from the late Cambrian, which would make them contemporaries of the early marine invertebrates. But, as Don Williamson points out, they could not have been the larvae of amphibians since the amphibians did not arrive onto the scene for another two hundred million years.

Some evolutionary biologists agree with the famous Oxford zoologist and biological "poet" Walter Garstang, who posited that evolutionary changes at the embryonic or larval stages were capable of changing the evolution of the adult form. From this perspective, chordates might have originated from ascidian tadpoles through larval features being carried into the adult form – an example of the mechanism that Garstang dubbed paedomorphosis. Building on Garstang's theory, E.J.W. Barrington, former Professor of Zoology at Nottingham University, has suggested that one possible way in which tadpole forms might have persisted into adulthood without metamorphosing was through failing to respond to environmental cues.[1] For instance, a tadpole larva that normally metamorphosed on contact with algae tethered to the ocean floor might have drifted or swum into deep water and simply failed to discover the metamorphic signal. But for this to lead to paedomorphosis, we would have to assume additional complex evolutionary developments, such as the capacity for reproduction at the tadpole stage. The tadpole might also have needed to feed – though, of course, the inability of the present-day tadpole to feed could be a later adaptation.[2]

According to Williamson, this rigmarole is unnecessary. In his view, the forerunner of the vertebrates was an adult marine chordate with tadpole features, rather like the Cambrian fossil tadpoles found in China, which hybridised with the ancestral urochordate to give rise to the sea squirt metamorphic life history. Williamson also suspects that hybridisation between a marine vertebrate with tadpole features and a nonmetamorphosing proto-amphibian might have given rise to the amphibian tadpole metamorphosis. While this might appear a bit of a stretch, the truth is that it would be difficult to propose any theory that does not stretch the imagination. For, as Barrington put it, 'the ascidian larva is a dual organism in which there is sharp demarcation between the temporary organs of the larva and the rudiments of the permanent organs of the adult'.[3]

The closer one examines this strange duality, the more bizarre it seems. Inside the fertilised sea squirt egg, the embryo develops, in conventional fashion, to the stage of a gastrula. 'From this point onward', in Barrington's words, 'two independent developmental mechanisms are operating side-by-side, the development of the larval structures proceeding virtually independently of that of the permanent [sea squirt] organisation'. Whatever the evolutionary origins for this remarkable situation, it must explain how the fertilised egg of a sea squirt contains not just a single distinct developmental programme but two, as well as how these two programmes, once they proceed past the common gastrula phase, are capable of developing simultaneously into two radically different life forms.

Presumably, very early in embryology, two stem cell populations must separate – one is reminded of the origins of identical twins in humans – but in this case each group of cells is programmed for independent development, with the tadpoles and juveniles growing side-by-side. Initially, the tadpole development proceeds at a more rapid pace, but the juvenile soon catches up. Meanwhile the very different larval and juvenile nervous systems also develop independently, whether side by side or one above the other, until eventually the tadpole settles to the ocean floor, its body degenerating to leave behind the juvenile adult. Moreover, though both the adult and the larva are bilaterally symmetrical, the developing juvenile follows a completely different body orientation. In the conventional evolutionary trajectory, one might expect the adult sea squirt brain to develop, through adaptation, from the larval ganglion, but in fact the adult develops an entirely new brain, and the tadpole brain is discarded.[4]

The presumed evolutionary relationship between these headless ascidians and the vertebrates suggests that we might look to the simpler forms to help us understand the evolution and developmental pathways of more complex life. But as long ago as 1968, Barrington urged that we do so with caution.[5] Sea squirts are relatively simple animals, and their metamorphosis is initiated at a very early stage in their development. The regulation of their development, including these dramatic changes, may not require any more complex mechanisms than would be found in the early embryonic development of a more complex animal. Nevertheless, it is interesting, in light of what we have learned about hormonal triggers in insect metamorphosis, to consider the commonalities in metamorphosis between marine invertebrates and chordates, from ascidian tadpoles to vertebrates. This perspective assumes, however much it strains credulity, that we humans are distantly related to sea squirts – a supposition suggested by the existence of the notochord and its accompanying nerve cord in tadpole larvae.

But there is also another, more subtle thread of causality that links the metamorphosis of the ascidians to chordate evolution, one that is still detectable today in our human brain.

We have already seen that marine invertebrate metamorphosis is primed to respond to environmental cues. Daniel Jackson and his colleagues at the University of Queensland have investigated the metamorphosis of the sea squirt *Herdmania curvata*, which lives on the underside of boulders and ledges on coral reefs, observing the dramatic transformation that begins when certain environmental cues come into contact with tiny sensory organs called papillae, which are located on the skin over the head of the tadpole larva.[6] These papillae send a nerve signal to local cells that release a key internal factor, which in turn triggers the developmental programmes that control metamorphosis. Other biologists have demonstrated that this internal factor is probably a protein, known as Hemps, which is localised to those same sensory papillae on the tadpole's head. Much as we have seen with ecdysone in insects, Hemps plays an important role in regulating developmental pathways, triggering a developmental cascade involving hundreds of genes, which programme the destruction of the tail and other unwanted larval organs and tissues, while simultaneously masterminding the development of the juvenile adult organs, tissues, and body form. All of this happens with amazing rapidity. Within minutes of the larva coming into contact with the environmental cue, retraction of the head papillae and resorption of the tail begin.[7] And here a more subtle developmental similarity with more advanced vertebrates comes into play.

In sea squirts, a duct-like organ known as the endostyle lies buried in the floor of the throat. It secretes a mucus, which is passed to the gut, in order to "fix" iodine – that is, it takes up free iodine from the ocean and chemically binds it to organic molecules. This process, called iodination, is a necessary step in forming thyroid hormones. Recent studies by Eleanora Patricolo and her research team in Italy have confirmed the presence of the thyroid hormone, thyroxine, in the tissues of sea squirt tadpole larvae.[8] Thyroid hormone also appears to play an important role in ascidian metamorphosis, so much so that when the same researchers administered a poison known to disrupt thyroid function, the larvae failed to metamorphose to the adult.[9] Other biologists have taken this a step further. The developmental gene *CiNR1* is linked to thyroxine activation. In 1998, Eleanora Carosa and her colleagues demonstrated that *CiNR1* was active in the sea squirt *Ciona intestinalis*.[10] In another primitive chordate, the lancelet, metamorphosis takes place through a larval stage, and its larva has an endostyle where iodine is fixed and where thyroid hormones have been discovered. These findings have led biologists to surmise that the endostyle may be the evolutionary forerunner of the vertebrate thyroid gland.

Thyroxine has two very different biological actions, one developmental and the other endocrine. The difference lies not in permutations of the hormone but in the place and timing of its action, whether during development or in adult physiology. Thyroxine acting hormonally in adult tissues works as a control lever on the rate of cellular chemistry, whereas thyroxine acting on embryonic or metamorphic tissues acts as a developmental master switch. When the hormone enters the nucleus, it discovers and binds to special receptors within the chromosomes, and this triggers key control genes to turn on developmental programmes. This pattern of regulation is vital to the way in which embryonic development proceeds. It is also vital to the genetic cascades that underpin metamorphosis. From an evolutionary point of view, it is interesting that the same subfamily of nuclear receptors, known as thyroid receptors or TRs, responds not only to thyroid hormones but also to the moulting hormone, ecdysone.

Thyroxine has been shown to play an important role in the development of sea urchins, crown-of-thorns starfish, and the sand dollar *Leodia sexiesperforata*.[11] Given the range of species involved, this suggests that development can be so conservative that it is very likely that interesting evolutionary links will be found.

Lampreys are primitive fish, some of which are free-feeding while others are parasitic on other fish. All lampreys have a cartilaginous skeleton that does not mineralise to bone. They are jawless and easily recognised from the sucker that surrounds their mouths and by the single nostril on the top of their heads.

Their skin is devoid of scales and slimy to the touch; their seven gill openings extend backwards from the line of their eyes. The parasitic varieties attach to other fish by means of the sucker, using their teeth-bearing tongues to rasp through the scales and extract their blood. Others feed on small invertebrates.

When it is time to spawn, even the marine varieties return to freshwater rivers and streams. After hatching, lampreys spend most of their lives in a larval phase, known as the ammocoete. This lasts from three to seven years, during which time they burrow into the soft sediments of stream and river bottoms and filter-feed on detritus. The metamorphosis is relatively slow, taking three or four months, during which time the animal undergoes extensive internal and external alteration, including the development of the eye, the loss of larval bile ducts and the gall bladder, the transformation of the lining cells of the intestine and gills, the total regression of the larval kidney and its replacement with an adult organ, and the development of teeth and tongue suitable for the adult life cycle. During the metamorphosis, the larval endostyle develops into a fully operational thyroid gland, with secretory follicles. The marine varieties of lampreys then return to the sea, where they spend a year or two in the adult phase before returning to the rivers to spawn, and usually to die.

The lamprey, though highly adapted to its modern lifestyle, is still widely regarded as a living representative of the early evolution of the vertebrates, so this transformation of larval endostyle to a thyroid gland during metamorphosis is particularly intriguing to developmental biologists. In the early decades of the twentieth century, scientists attempted to trigger lamprey metamorphosis with thyroid extract, but all such efforts failed, suggesting that thyroid hormones played no part in the process. In fact this interpretation was quite wrong – for an interesting reason. In lampreys, thyroid hormone acts like juvenile hormone in insects, inhibiting rather than stimulating metamorphosis. In bony fish, such as eels and flounder, the situation is the exact opposite, so that treatment with thyroid hormones induces precocious metamorphosis.[12] As long ago as 1928, E. Murr and A. Sklower studied six arbitrary stages in eel metamorphosis, during which they found that the volume of the thyroid gland increased some fourteen fold during the early stages. Since then, many studies have confirmed the role of thyroxine and its pituitary-driven negative feedback control, through thyroid stimulating hormone, in the metamorphosis of a wide variety of bony fish, including eels, flounder, flatfish, zebrafish, sea bream, coral trout grouper, and the orange-spotted grouper.[13] These observations may have applications for breeding in commercial fish farms. More importantly, they have evolutionary ramifications, since fish occupy a vital step in vertebrate evolution.

At some time in the distant past, a fish-like ancestor emerged onto land to give rise to the first four-legged animals, or tetrapods, which in turn evolved to the terrestrial vertebrates. The next stage in this complex, many-branching, and sometimes reuniting evolutionary trail are the amphibians.

*African clawed frog*

## 24

# Of Frogs and Their Relatives

THE LIFE CYCLE OF FROGS and toads, with their meta-morphosis through a tadpole stage, is every bit as enchanting to schoolchildren as the quasi-magical transformation of cater-pillars to butterflies. In fact, the closer we look at these famil-iar amphibians, so similar in form to ourselves that they have become the stuff of fairytales, the more interesting their real life story becomes. They truly are amazing creatures.

The first revelation, when we examine the evolutionary ori-gins of frogs, is to find that they are so exceedingly ancient. Amphibians are thought to have evolved from a group of lobe-finned fish, the rhipidistians. These gulped air and dragged themselves through the mud at the water's edge using bony fins. In time, the fins evolved to the four limbs of the first land-living vertebrate animals. Spectacular fossils of this evolu-tionary step from the oceans to the land were discovered on

Ellesmere Island, in Canada, in 2004. They document the existence, approximately 375 million years ago, of a fish-like animal, now called *Tiktaalik*, that grew to between four and ten feet in length. *Tiktaalik* appears to be an intermediate species, between fish and reptiles, displaying a typically fishy body shape, jaws, fins, and scales but the head, neck, and ribs of a land animal. Within the fins, the bones have also evolved to an intermediate stage to support both swimming and terrestrial weight-bearing.[1] These pioneering vertebrates did not always share our familiar five digits on the ends of their four limbs. Three different genera of early animals have now been discovered – *Tulerpeton, Ichthyostega,* and *Acanthostega* – all from the late Devonian period, some 390 to 340 million years ago, that have, respectively, six, seven, and eight toes.[2] The oldest reptile fossils are almost as ancient as those of the amphibians, suggesting that the amphibians and the amniocetes – those with hard eggs, including early reptiles, birds, and mammals – quickly diverged from a common ancestor, perhaps through differing adaptations to amphibian and solely terrestrial life cycles. The earliest frog fossils, found in North and South America, date to the early Jurassic period, about 208 million years ago, which makes them contemporaries of the early dinosaurs. But where the dinosaurs, other than those that evolved into birds, became extinct, frogs flourished.

Today's frogs have diversified to more than four thousand estimated species and they have extended the range of ecosystems they inhabit to every continent other than Antarctica. By any measure, this is an extraordinary evolutionary success story. It simply begs an explanation.

Of course, frogs are easily recognisable, the small, tailless animals with a squat body and long, folded hind legs adapted to jumping. They have large bulging eyes and moist skin. Typically they live on land or in trees close to water. The adults come equipped with lungs, which they use to breathe air when out of the water. Surprisingly, when they are under water, they can also breathe through their skin, which is richly supplied with blood vessels that allow oxygen to diffuse through it and enter the blood. Toads, which are members of the same order of amphibians as frogs, are toothless, have a dry, warty skin, and are adapted to survive in drier ecologies. In spite of their ancient colonisation of land, the great majority of frogs lay eggs in or around water so that their larvae, tadpoles, hatch out in this ancestral habitat. Amphibian tadpoles have no limbs, breathe through fish-like gills, and swim with a powerful fish-like tail. This dual existence of developmental forms in one life cycle, adapted first to water and then to land, is characteristic of the class Amphibia, which also includes newts, salamanders, and caecilians, and distinguishes it from the class Reptilia, which includes turtles, snakes, crocodiles, and alligators, not to mention the now extinct dinosaurs.

The four thousand different species of frogs range from the Brazilian gold frog, just one centimetre long and weighing less than thirty grams, to the West African goliath, which measures thirty centimetres or more and tops the scales at three kilograms. Frogs' jumping locomotion enables them to leap up to twenty times their body length, a remarkable feat that puts our Olympic long jumpers into the shade. Like the capacity to live and breathe both in water and on land, one can only imagine that this impressive adaptation must have helped them escape in times of danger. To accommodate the leap, frogs have evolved very long legs in relation to their body size and a specialised pelvic girdle designed for power and shock absorption.

It will come as no surprise to learn that there are many variations of development among the amphibians, and even within the large and diverse order, the Anura – the frogs and toads and their familiar tadpoles – there are exceptions to the metamorphic life cycle. In some species, the female gives birth to tiny, fully-formed frogs, the tadpole phase subsumed within the mother prior to birth. Nevertheless, metamorphosis from tadpole to frog is more typical.

According to James Norman Dent, based at the Department of Biology at the University of Virginia, Charlottesville, 'a remarkable series of changes takes place when an aquatic fish-like tadpole is transformed into a land-dwelling frog'.[3] These embrace the massively destructive in terms of larval tissues and organs and the hugely creative in terms of the construction of the new adult. Some of the changes arise slowly and progressively within the lifetime of the tadpole, while others take place very rapidly, while the larva is still in the water, at the time of metamorphosis.

Metamorphosis begins with the formation of a membrane, the operculum, that grows over and covers the tadpole's gills. The head, meanwhile, is extensively remodelled as the larval cartilage is remolded to the adult bony cranium, and the horny anuran beak begins to form. New colour patterns invade the skin. The front and back legs, which had already begun to develop during larval life, rapidly elongate, and the tail is reabsorbed, causing the anus to move into a different position. Inside the body, there are dramatic changes to the nervous system and eyes, wholesale reconstruction of the great arteries from the heart, shortening of the gut, degeneration of the larval skin and replacement with the adult skin, and formation from scratch of the hyoid cartilages in the throat, which, together with growth of the chest muscles, make it possible for the adult to breathe air.

The changes extend to the molecular level. Genes expressed in the liver are reprogammed, and those responsible for types of haemoglobin are switched. There are also changes in the expression of the gene responsible for the hardening of skin, and significant changes to the immune system.[4] As we have seen in other varieties of metamorphosis, these genetic cascades bring about the massive destruction of larval tissues and organs, concluding with death and dissolution of unwanted cells and wholesale reprogramming and regeneration at every level, from genes, through to cells through to tissues and organs.[5]

Inevitably, scientists have searched for a role of the thyroid gland in these catastrophic algorithms of destruction and reconstruction. Thyroid extract was found to hasten frog metamorphosis as early as 1912. Over subsequent decades, it became clear that thyroid hormone is fundamental to the entire process, activating key control genes that initiate the genetic cascades of metamorphosis. And just like ecdysone in insects, the production of thyroid hormone is subject to negative feedback in response to the pituitary gland. This feedback system kicks in during the early larval stages of the amphibian life cycle.[6] We humans share the same pituitary-thyroid axis. Like frogs, our pituitary gland is linked to the hypothalamus, in the midbrain.

In response to some external signal, which is conveyed to the hypothalamus, the developing amphibian pituitary gland secretes thyroid-stimulating hormone, TSH, which is carried through the blood stream to the thyroid gland, where its presence activates the release of thyroid hormone. This, in turn, not only triggers metamorphosis but also, through negative feedback, switches off the release of further TSH. Moreover, the body structures, internal organs, and biochemistry of amphibians are a good deal more complex than those of insects and, for the complete panoply of metamorphosis to take place, other hormones are also important – including prolactin, adrenal steroid hormones, and melatonin, a hormone that controls skin pigmentation and which responds to prevailing light and dark conditions, possibly cued by the duration and intensity of light falling onto the eye. The duration and cycles of daylight versus night have been shown to influence the speed of metamorphosis, with observable effects not only on the rate of change but even the rhythms of cell proliferation in the skin. In the opinion of Jane C. Kaltenbach of Mount Holyoke College in

Massachusetts, 'it is possible that the effects of lighting changes ... are mediated by the pineal gland and [the hormone] melatonin and are partially responsible for the "occurrence" of spontaneous metamorphosis when conditions in nature are favourable'. The pineal gland is adjacent to, and intimately connected with, the pituitary, which would allow environmental cues to trigger hormonal secretion. Thus, signals arising from low environmental temperatures have also been shown to slow down, and perhaps even inhibit, metamorphosis.

Donald D. Brown, Adjunct Professor in the Department of Embryology at the Carnegie Institution of Washington, DC, has for many years been researching the way in which thyroid hormone controls metamorphosis in amphibians. In particular, he has studied metamorphosis in the African clawed frog, *Xenopus laevis*, a plump, medium-sized aquatic frog with smooth, slippery skin, webbed rear feet, and the eponymous clawed front feet.[7] In a series of experiments, Brown and his colleagues have introduced mutant genes into the sperm of the frog to create an abnormal thyroid hormone receptor protein in the offspring. This competes with the normal receptor during metamorphosis, when the mutant receptor blocks the normal developmental response to thyroxine. The result is a catastrophic failure of every metamorphic change in the tadpoles. From such experiments, they have concluded that metamorphosis in frogs and toads follows a set of complex developmental programmes in which the entire process is controlled by thyroid hormone.

We have already seen how, during amphibian metamorphosis, tissues and organs grow *de novo*, die, or are remodelled. The experiments of Brown and other researchers have shown that thyroxine works by changing the expression of genes in those tissues that are remodelled or die during metamorphosis. Moreover, it has been shown that thyroxine achieves this effect in two ways: directly at the level of the single cell, where it controls cell differentiation into specific types necessary for tissue and organ development, and also at the level of cell-to-cell signalling within tissues and between adjacent tissues. This is another important step in understanding not merely amphibian metamorphosis, but development throughout all of multicellular life.

Vincent B. Wigglesworth didn't live long enough to witness the first unfolding of the human genome. I have no doubt that he would have found it fascinating. For me, as for many biologists, one of the most revealing aspects was how much genetic programming we have in common with every other form of life on Earth. For example, we share 2758 of our genes with the fruit fly and 2031 with the nematode worm; and all three of us – human, fly, and worm – have 1523 genes in common.

Even more pertinent is the common evolution of Hox genes, which help to regulate embryonic and post-embryonic development along the front-to-back axis of animals. So important are the Hox genes that they are now seen as 'a guiding force within the field of evolutionary developmental biology'.[8] Hox genes are exceedingly ancient and have been remarkably conserved from marine invertebrates to flies and humans. To put it in the words of David E.K. Ferrier and Carolina Minguillón, 'The finding that galvanised the EvoDevo community was that in the Hox genes we have [similar] genes acting in [much the same way] across the animal kingdom'.[9]

Common genes, common developmental control mechanisms – perhaps it is not altogether surprising that essential commonalities should extend from the developmental mechanics of metamorphosis in marine invertebrates, through primitive chordates, through marine vertebrates, through the common ancestor of amphibians, reptiles, and mammals, all the way to us, humans.

*Human foetus*

# 25

# To Be Human

'MAN', TO QUOTE Jacob Bronowski, 'is a singular creature. He has a set of gifts which make him unique among the animals: so that, unlike them, he is not a figure in the landscape – he is a shaper of the landscape'.[1] What makes us quintessentially human is not the fact we can walk on two legs – birds do – or the fact we have exquisitely dexterous limbs – octopuses have eight of these. We are human because we, alone of all the species that evolved on Earth, have the gift of sentience. How then can we come to understand how our singular species belongs to the same natural world as does a frog or a honey bee, and yet from out of the complexity of nature's history we have become, in Bronowski's words, 'explorers of that nature'?

Early in the evolution of terrestrial vertebrates, the lines leading to the amphibians split off from the common ancestral line that led first to the reptiles and later to the mammals. At this key point of divergence, the metamorphic inheritance underwent a fundamental schism. The amphibian line, in bridging the two very different ecologies of water and land, evolved a new developmental drama of metamorphosis; meanwhile, the reptile-mammalian line abandoned it.

The lengthy development of the foetus within the mammalian womb allows more time for complex embryonic development than was possible for our egg-laying ancestors, and this has subsumed the traces of any metamorphic inheritance. But, though we cannot see it from the outside, the quasi miracle of hormone-controlled development still remains. The pregnant mother can sense some of the changes taking place inside her. She feels it directly when foetal hormones affect her body's chemistry and she thrills to the first tiny movements of the developing foetus. And, looking back over our shoulders at the mammals as a class, here and there we glimpse reminders of that dramatic metamorphic past.

Metamorphosis is now viewed as a new phase of post-embryonic development and, as we saw earlier, humans exhibit a singularly important example of this type of development at the time of puberty, suggesting to some biologists that puberty in humans is a form of metamorphosis.[2] Much as metamorphosis in insects and amphibians is brought about by signals that begin in the central nervous system, leading to the release of ecdysone and thyroid hormone, puberty in humans is also brought about by pulses of gonadotropin-releasing hormone, which is released by the hypothalamus. This stimulates the neighbouring pituitary gland to increase the secretion and release of the sex-gland-stimulating hormones, or gonadotropins, which travel trough the blood stream to the ovaries and testes, where they stimulate increasing blood levels of oestrogens and androgens. And just as environmental cues can trigger metamorphosis, environmental factors can also influence the timing of puberty. For example, improved nutrition may have lowered the age of female puberty during the twentieth century; likewise, environmental contamination from bisphenol A, a common component in plastics production, may hasten the onset of puberty.[3] Puberty, as well as human fertility, can also be considerably influenced by physical illness and environmental stress, the latter almost certainly mediated by input through the hypothalamic link, echoing the environmental signalling we saw in amphibian reproduction.[4]

Moreover, puberty is not the only period of post-embryonic development in humans. Unlike most other animals, humans retain the embryonic potential for brain development long after birth, a facility that allows our exceptionally large and complex brains to grow and develop throughout the first two years of infant life.[5]

The human body contains about two hundred different types of cell, specially adapted for the different tissues and organs. Almost all of these different cells, tissues, and organs are formed during embryonic development. As Tim Mohun and Jim Smith of the National Institute for Medical Research in London explain, we can observe a similar specialisation in frog spawn with little more than the assistance of a magnifying glass.[6] The embryo changes rapidly and accurately from the ball of cells known as the blastula to more complex shapes containing the first evidence of distinct tissues, from brains to limbs. Research on how such developing cells recognise their own tissue type is the basis of the present wave of stem cell research that may soon help doctors treat serious illnesses, such as Parkinson's disease and the damaging effects of heart attacks.

Looked at more broadly, it's clear that the entire developmental plan of a human being must be contained within the egg immediately after fertilisation. But this does not mean that every cell derives from an autocratic central organisation, with no room for variation. We see this potential for downstream changes in the slight differences that appear in identical twins, which are brought about through epigenetic control over gene activation as opposed to genetic variation – an astonishing additional layer of environmentally sensitive influence over our genes, both within the womb and throughout our lives. Indeed, those same epigenetic control systems are critical in deciding the fate of embryonic stem cells: whether they will become liver, blood, or brain.

So it is that with modern advances in developmental biology, we realise that embryonic development is determined by a combination of mechanisms: the central "genome-determined" plan, the critically important epigenetic systems of gene control, and an additional series of complex interactions between adjacent cells and tissues. All of these potentials are pre-programmed into the human egg from the moment of its fertilisation.

The idea that every individual human being arises from a single pluripotent cell never ceases to amaze me. We have evolved from humble beginnings, much as any other life form, and we develop from a fertilised egg along similar developmental principles to other animals. And like all multicellular animals, plants, fungi, and even many of the single or few-celled life forms known as protists, we are composite beings. We breathe oxygen thanks to the mitochondria that originated at least a billion years ago from the symbiotic incorporation into our ancestral cells of free-living, oxygen-breathing bacteria. A significant part of our genetic inheritance came into being from the progressive incorporation of large numbers of retroviruses, similar in genetic structure to HIV-1, the cause of AIDS, which, like the bacteria that became mitochondria, have long since been incorporated into our normal genetics and development.[7] At least eight of these endogenous retroviruses appear to play

important roles in normal human reproduction and the formation of the placenta.[8] We are dependent on symbiotic bacteria
in our guts for healthy digestion and on myriad plants and other
animals for the essential vitamins, fatty acids, and amino acids
that keep us healthy from day to day. And still, the connections
that integrate us into the biosphere of our beautiful ocean-dom-
inated planet extend deeper: to the most profound levels of our
biochemistry, our physiology, and the genes and chromosomes
that are responsible for our programming and inheritance and
for the cellular structure and machinery of development that
enables the wonder of every individual human life.

On day one of your origins, the first cleavage of that fertilised
cell took place. You were now two cloned cells, each inheriting
matching pairs of twenty-two chromosomes from each of your
parents in addition to the two sex-determining chromosomes,
making up the human chromosome complement of forty-six.
If, in addition to the single X chromosome from your mother,
you inherited an X chromosome from your father, you are now
female; if a Y chromosome, you are now male. By day four, you
had become sixteen, then thirty-two cells, a solid planula-like
ball called a morula. The following day, the morula developed
into the familiar hollow embryo of a blastula, with an outer
wall of cells and an inner cell mass surrounding a fluid-filled
core. You subsequently developed from that inner cell mass,
or embryonic disc, and over the following couple of days your
blastula, now technically a blastocyst, attached to the lining
of your mother's womb, where it burrowed in and implanted
itself, making itself a home. At the area of contact with the
womb, your endogenous retroviruses controlled the cellular
differentiation of the placental interface, fusing the cells into
a confluent monolayer; at the same time, they helped to make
the human placenta, the deepest, most invasive, and highly
specialised reproductive interface of all the mammals, shared,
along with the contributing retroviruses, only with the great

apes. Your blastocyst began to secrete the hormone human chorionic gonadotropin, or HCG – the chemical basis of the popular at-home pregnancy tests – which entered your mother's circulation and stimulated her ovaries to produce oestrogen and progesterone to prevent menstruation.

It did not trouble the emerging you that it also made her liable to morning sickness. Over days seven to ten, the indentation of a blastopore pushed into your blastula wall, signalling that you were turning into a gastrula. Up to this point, each of your cells was still a clone of that first fertilised cell. But with gastrulation you underwent the first developmental change that would differentiate your cells into the three fundamental tissues that would fashion all of your adult tissues and organs: the ectoderm that would become your skin and nervous system; the endoderm that would form the lining of your gut and your internal organs; and the mesoderm that would become your muscles, bones, and heart. Your blastopore would not become your mouth, but the opening at the latter end of your intestinal tract, your anus. Your future mouth would have to break through to the surface ectoderm at the other end, making you, like all other members of the chordate lineage, a deuterostome. The shared inheritance runs even deeper. Your abdominal lining, or coelom, would develop from outpouchings of your primitive gut, making you an "enterocoelous deuterostome".

By days fifteen to twenty-one, your developmental inheritance, and in particular your very ancient Hox developmental genes – quadruplicated as a result of two rounds of whole-genome duplication, which were probably the result of two major hybridisation events long ago – began to control development along the length of your body.[9] Your body form assumed a typical bilaterally symmetrical shape, with the front end taking on the first indications of a head and the rear end taking the form of a tail, with two parallel rills appearing along your front-to-back axis, heralding the "neural groove" that would fold over into a cylinder to form your future spinal cord and, at the head end, the brain that is currently directing the reading of this book. By day twenty-one, you had vestigial gills, Haeckelian relics of a marine past, together with a more pronounced stubby tail. From weeks four to eight, all the major organs of your body were formed. As early as twelve weeks, most of your organs and body parts were developed, with the exception of your brain and lungs. From this point onwards, the bulk of your organs and tissues were now developed, though they still needed to grow – with the exception of your brain, which was still largely undeveloped.

The extraordinary thing, yet a thing that is absolutely obvious when we think about it, is that when you first arrived into the world, as a newborn baby, your brain was still relatively small – little more than a quarter of its present size – and much of its cortical, or higher, functions had not been developed. During the early days after birth, your brain added approximately a quarter of a million nerve cells every minute to its mass and this colossal increase in brain size and complexity continued during your first two years. If we compare this to our closest evolutionary relatives, the chimpanzees, the ratio of brain weight to total body weight is the same in both species at birth. But such is the post-embryonic growth of the human brain that by the time we are adults, our brain-to-body-weight ratio is three and a half times that of the chimpanzee.[10] This oversized brain works through connections, known as synapses, between nerve cells. Roughly thirty thousand synapses per square centimetre of brain cortex are formed per second during the first few years of life.[11]

It is this prodigious development of our brain, during both our embryonic and post-embryonic stages, that makes us quintessentially human. And here, too, we discover a telling link to the metamorphosis story.

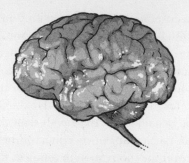

*Human brain*

# 26

# *The Sentient Brain*

IN 2004, IN AN EDITORIAL published by the Endocrine Society, R. Thomas Zoeller of the Biology Department at the University of Massachusetts at Amherst made an intriguing statement. 'Insights gained from the studies of frog metamorphosis', he wrote, 'are helping us understand the role of thyroid hormone in the development of a completely different tissue – the human brain'.[1] Such a statement should, of course, cause readers to sit up and take notice.

Zoeller did not start out his academic life with an interest in developmental biology. After a spell in the US Army, from 1972 until 1974, he spent his last weeks working alongside a fish and wildlife manager, which in Zoeller inspired an interest in biology and ecology. He decided he would enrol as an undergraduate at Indiana University, where a group of scientists, including Robert W. Briggs, were studying amphibian development using the African clawed frog, *Xenopus laevis*, as an experimental subject. As a graduate student, Zoeller moved to Oregon State University, where he was encouraged by the work and personality of Frank Moore, a young and enthusiastic professor who was fascinated with the hormonal control of behaviour. For his PhD, Zoeller studied the physiological basis of reproductive behaviour in the rough-skinned newt, *Taricha granulosa*, and in particular the coordination between environmental factors and the motivation to reproduce.

To put it simply, the rough skinned newt would only engage in sexual behaviour in certain circumstances. In particular, the male newt would only do so in the presence of a sexually competent female. This may sound obvious, but let's take a mental step backwards and think about what's going on within the newt brain and endocrine system: there must be visual signals that tell the male that a female is nearby, and perhaps also pheremonal signals that tell him she is reproductively competent. As we have seen in so many examples of metamorphosis, such external cues activate internal responses, in this case biochemical and very likely hormonal ones. Zoeller examined minute samples of the newt brain, looking for any chemical responses that would help him understand the way in which this complex behavioural system actually works.

In time he found some changes in the levels of certain chemicals, known as neuropeptides, and these looked as though they exhibited a significant response. He pushed on, becoming interested in the role of the neuropeptide arginine-vasotocin, which is involved in sleep regulation. But in the end he was frustrated. In his PhD dissertation, he concluded that merely looking at changes in peptide content could not tell him enough to really understand the system.

In 1984, having finished his doctorate, he moved to the US National Institutes of Health, in Bethesda, Maryland, where he studied the nerve cells that control the hormonal system of reproduction. Entomologists such as Wigglesworth and Williams had shown that an external cue sends a message to key cells in the insect brain, which then initiates the internal and behavioural responses of metamorphosis. In vertebrates, including frogs as well as humans, a distinct population of brain cells, the gonadotropin-release-hormone, or GnRH, cells, control reproduction. Scientists at NIH had found a way of identifying GnRH within these brain cells, which allowed them to study the regulation of sex hormones in reproduction. Zoeller heard on the grapevine that Peter Seeburg, then working at the biotechnology company Genentech, had cloned the rat gonadotropin-release-hormone gene. At much the same time, the chemical structure of thyrotropin-releasing hormone, or TRH, was published. TRH regulates the release of thyroid hormone through control of the pituitary-thyroid hormonal axis.

With this in mind, Zoeller began to investigate some of these nerve cells, or neurons, in the brain. He began working on the cells that make GnRH, but he soon ran into a problem. The cells that make the hormone are not gathered together in a distinct nucleus, but are scattered throughout a region of the hypothalamus. In the rat brain, for instance, there may be two thousand GnRH nerve cells spread diffusely over a cubic centimetre. In microscopic terms, this was a very big area, especially so when you were looking at one cell at a time, which made it difficult to work with the GnRH system. So in 1985, wholly for pragmatic reasons, Zoeller shifted his focus to the thyrotropin-release system, since the bodies of the nerve cells that make TRH happen to be focused in a single compact collection, known as the paraventricular nucleus. From here, their thread-like nerve strands, carrying nerve signals and hormones, travel out of the nucleus to enter the hypothalamus. As we have touched on already, the hypothalamus is one of the most important structures in the vertebrate mid-brain, playing a central role in the regulation of the master gland, the pituitary. We might also recall that it is a critical part of the tadpole brain during amphibian metamorphosis. All of the brain-derived tropic hormones, such as the gonadotropin- and thyrotropin-releasing hormones, converge on the hypothalamus, where a system of blood vessels ferries them to the pituitary.

Until 1986, Zoeller had thought only dimly about embryonic development, and little or not at all about the role of thyroid hormone in development. Then his son was born. In his words, 'The birth of a child should change you. If it doesn't change you at the moment of birth, it's gonna change you later.' It certainly changed Zoeller's interpretation of what was really important. Two years later, he moved to the University of Missouri School of Medicine, where he worked on two projects: the negative feedback system of thyroid hormone on neuronal levels of TRH; and the effects of alcohol on this same system.

At the time, it was known that deficiency of thyroid hormone in the human foetus could lead to severe mental retardation as well as to difficulties with movement and coordination in late life. Yet, getting a full understanding of the underlying mechanisms had been hampered. Many scientists believed that the placenta broke down maternal thyroid hormone, preventing it from entering the foetal tissues, and so doctors assumed that thyroid hormone had no effects on brain development until after birth, when the infant's own thyroid gland became functional. But a number of investigations had recently thrown doubt on this. Starting in 1989, several researchers had shown that thyroid hormone of maternal origin appeared to cross the placenta and enter the foetus.[2] If thyroxin was important in brain development, this made sense, since most of the key developments in the mammalian brain took place before birth. Around the same time, other researchers had discovered something rather puzzling: the levels of the activated thyroid hormone, known as triiodothyronine or T3, in the developing cerebral cortex of the human foetus were much higher than would be expected from the circulating levels of the hormone in the foetal blood.[3] This made Zoeller sit back and think again. 'The more I thought about [it], the more I realised that the time when thyroid hormone exerts a very profound effect on the brain must be during [embryonic] development'.

In 1992, he read an important paper by Scott Young and his colleagues, in which they probed the foetal brain searching for developmental genes that might be targeted by thyroid hormone.[4] For a hormone to target a developmental gene, there must be a hormone receptor site in the chromosomes close to the gene. There are two main types of thyroid hormone receptor in the brain, alpha and beta receptors. Young and his colleagues were able to map the genetic expression of these two thyroid receptors in mammalian brain cells in terms both of their spatial distribution and their expression over time. Zoeller was intrigued by the results: an exquisitely outlined pattern and complexity of thyroid-receptor expression, which must surely be the key to what thyroid hormone was really doing during foetal development.

Zoeller began to design a new series of experiments to test this further. But it wasn't until he made a further move, to his current base at the University of Massachusetts at Amherst, that he was in a position to conduct the first one. If, as he now suspected, thyroid hormone plays a pivotal role in mammalian brain development, specific genes within these developing brain cells must be critically responsive to the presence of thyroid hormone. He needed to find a way to identify those genes.

He knew that other scientists had looked for the same genes and failed. 'I thought the reason they failed was that they were looking at the differences between foetal brain development in a normal pregnant animal and an animal that had been deprived of thyroid hormone for a long time'. He needed a different methodology. Instead of comparing normal maternal rats to thyroid-hormone-deprived rats, he began with a test group of pregnant rats, known as dams, that were hypothyroid to start with. He then timed his experiment for a very early stage in pregnancy, when the foetus would not yet have developed its own thyroid function and thus foetal brain development might be dependent on the maternal hormone crossing the placenta into the foetus. He compared a control group of hypothyroid dams with a hypothyroid group that had been given a single subcutaneous injection of thyroxine. Using this new methodology, Amy Dowling, a graduate student working under Zoeller's supervision, found eleven genes that were selectively affected by thyroid hormone, seven of which appeared to be enhanced by thyroxine and four to be suppressed by it. She went on to demonstrate that at least two of these genes coded for proteins that played an important part in brain development.

This was an important discovery, the first time in the scientific literature that thyroid hormone was shown to regulate genes in the developing mammalian foetal brain.[5] It had clinical significance in suggesting that doctors wishing to avoid damage to the foetus needed to treat hypothyroid mothers, or potential mothers, as early as possible, before or during pregnancy. It also left an important question unanswered – a question that linked this work to evolutionary biology and what had been learned about development from the study of metamorphosis. What, Zoeller now asked himself, was thyroid hormone really doing to the developing human brain?

The forebrain – what we commonly refer to as our cerebrum, or cerebral hemispheres – is central to human intelligence, morality, and sentience. The cerebral hemispheres are wrapped around big, fluid-filled chambers, known as the lateral ventricles, and the developmental centres that construct the embryo's forebrain lie close to these chambers. While they are in the developmental centres, developing neurons undergo eleven rounds of reproductive division, then migrate outwards, towards the convoluted sheet of brain tissue we call the cerebral cortex, eventually populating the entire cortex in a series of six sequential layers. When the cells first proliferate, they express the beta-type thyroid-hormone receptor in their genomes. But as they migrate further outwards in the forebrain, the beta receptor is turned off and the alpha receptor is turned on. To Zoeller, that switch has to mean something. In addition, the number of reproductive divisions in the cells developing deep in the brain is not affected by thyroid hormone. Yet, when these cells begin to migrate, it seems they are already committed to their eventual fate, whether to become a functional neuron or a connective tissue support cell, called a glia. Thus, the absolute number of nerve cells that eventually populate the cerebral hemispheres is set at the earliest stages of development, which must have important implications for brain function in the future individual.

But what exactly decides each cell's fate? Zoeller's research suggests that the thyroid beta receptor, which is turned on at this stage, is the determining factor. Once the cell's fate is decided, the beta receptor is switched off and the alpha receptor is switched on, to play a role that has yet to be determined.

If thyroid hormone, through interaction with its various receptors, is playing such an important role in human brain development, how, at the molecular level, is it doing so? For Zoeller, the answer may come from scientists such as Donald Brown and their work on metamorphosis in amphibians.

We have seen how metamorphosis in the sea squirt tadpole, bony fish, and amphibian is controlled by thyroid hormone. In the African clawed frog, not only does thyroid hormone control the metamorphosis from the tadpole to the adult but, as Zoeller points out, metamorphosis takes the form of an orderly sequence of events, so that different tissues undergo thyroid-hormone-dependent metamorphic changes at different times and at different rates, in spite of the fact there is always a similar level of thyroid hormone in the blood. This suggests that there must be some system of control at the cellular level independent of the levels of hormone in the blood. It also suggests that thyroid hormone might have some peculiar property when compared to most other hormones – a property that might explain why it has remained such a key control factor in the development of chordates across the vastness of evolutionary time.

This line of thinking was borne out when, in the 1990s, Kathryn Becker and her colleagues demonstrated that local intracellular level of active thyroid hormone, T3, did indeed vary, even when the blood levels remained the same, and that such local intracellular variations were controlled by the actions of two opposing enzymes, one of which generated active T3 and the other which degraded active T3.[6] The counterposing actions of these enzymes made possible the thyroid hormone's binary switching system of control within tissues. This simple but elegant system also offered the means of controlling the timing of the action of the hormone during development. In the African clawed frog, for example, metamorphosis from tadpole to frog involved major changes in the positioning of the eyes and in the visual fields perceived by the eyes. These were accompanied by the formation of a new nerve projection from the retina to the brain. Don Brown and his colleagues had shown that these dramatic visual and neurological developments are controlled by the opposing enzymes.[7]

Just like frog metamorphosis, embryonic development of the mammalian brain takes place in an orderly sequence of events, and the timing of these maturational events is similar across a variety of mammalian species.[8] In 2004, Monique Kester and her colleagues reported finding the same binary system of control of local thyroid hormone in the cells of the foetal human brain during development, notably in the cerebral cortex.[9] As in amphibian metamorphosis, the amount of active hormone in the developing human foetal brain derived not from the circulating blood levels of thyroid hormones, which were much lower than in the adult, but from differential activity of the opposing enzyme systems within the cells. T3 levels increased in the developing cerebral cortex between weeks thirteen and twenty, until they reached levels higher than those found in the adult cortex. In contrast, T3 levels in the cerebellum, a part of the brain tucked under the cerebral hemispheres that is responsible for balance, were very low at this time and increased only after mid-gestation, at twenty weeks. A similar system managing local bioavailability of T3 was found to operate in the brain stem, the spinal cord, the basal ganglia, and the mid-brain region known as the hippocampus.

'Would you say that thyroid hormone is essential not for the development of the brain *per se*,' I asked Zoller, 'but for the development of a *normal* brain?'

'Yes, that is true. A good example is one of the genes that Amy pulled out – it's a gene called HES, which stands for "hairy enhancer of split". This is a gene that essentially blocks neurogenesis – the developmental centres' creation of nerve cells. 'Now it doesn't block proliferation but it blocks the differentiation of an undifferentiated cell to a neural lineage. Instead these cells become glia.'

'That's a terrific realisation, isn't it?'

'Yes it is. … We're kind of making a proposition – it's a very difficult thing to study in vivo – but we're proposing that thyroid hormone controls the balance of the early production of nerve cells and glia. And that to me suggests that without thyroid hormones, you have the same number of cells as before: it's just that the balance between the two kinds of cell is lost.'

To put it into simple words, Zoeller and these other researchers have discovered that this system, involving the opposing enzyme systems that control intracellular levels of active thyroid hormone – a system that evolved as a master control mechanism throughout hundreds of millions of years of vertebrate metamorphosis – also controls the absolute numbers of nerve cells that go on to form the cerebral hemispheres of the human foetus. It seems inevitable that thyroid hormone played an important role in human evolution, and especially the evolution and continuing development of the brain.

Thus, in metamorphosis we are witnessing the wonder of development. From the fertilised egg to a complex human adult, we witness that same wonder. And now that we grasp the commonalities, our vision opens onto a window of more universal understanding of the very wonder of life, in all the serendipity and linkages of the remarkable evolutionary story – its beauty inextricably one with its mysteries.

# Part V

# The Neverending Story

*Evening has overtaken me, and the sun has dipped below the horizon of the Ocean, yet I have not had time to tell you of all the things that have evolved into new forms.*

Ovid,
'Metamorphosis'

*Burgessia bella*

## Epilogue

# A Sting in the Tail

VINCENT B. WIGGLESWORTH'S contribution to our understanding of metamorphosis shines like a beacon through the pages of this book. This work was largely completed while he was working at the London School of Hygiene and Tropical Medicine, prior to his moving to Cambridge, to take up the Quick Professorship of Biology. But his wider contribution to insect physiology had only just begun. In a lifetime that began in 1899, when Queen Victoria was still on the throne, and which spanned almost the entirety of the twentieth century, sixty years of this spent in productive research, the great entomologist radically changed the way we think about the life of insects, their development, and – more broadly – the physiology and development of life.

He was the 'father of insect physiology', according to one of his students, Michael Locke, now Professor and Chairman of Zoology at the University of Western Ontario, who in 1995 wrote an obituary of Wigglesworth for the Royal Society in London.[1] In correspondence with Locke, I asked him if he thought this was a reasonable appellation.

'My authority for this', he replied, 'would be Carroll Williams. He told me that when the volume [*The Principles of Insect Physiology*, published in 1939] first came out, he was sitting with colleagues in a discussion group formed with the express purpose of going through the book, and they were awed by what they read. It was what they had been waiting for – a book that systematised insect physiological knowledge to create a coherent science'.[2] It was for this reason that, in the same year, Wigglesworth was elected to fellowship of the Royal Society.

At Cambridge, where he spent the final two-thirds of his career, Wigglesworth's genius was to lift the study of insect development beyond any applied scientific importance, admirable though this might be, to the level of a more universal understanding. To such a scientist, his ideals, the inner forces that drove and inspired him, would appear every bit as noble as the creation of the finest work of literature, or any other of the arts. John S. Edwards of the University of Washington's Zoology Department tells of how a single Wigglesworth article influenced him: 'In a small colonial Zoology Department where dusty vertebrate comparative anatomy reigned, my introduction to the seeming simplicity and tractability of the insect, as revealed by Wigglesworth's paper, with its clear, direct prose and elegant spare line drawings, was the conversion experience'.[3] He then quotes the American poet laureate Robert Frost, who said that a poem should begin with delight and end in wisdom. 'On that basis it might be claimed that many of Wigglesworth's best papers were poems of development'.

For some, Wigglesworth's approach has come to represent a lost purity of principle and endeavour. In April 1997, Michael Locke noted despairingly how modern science had entered a perilous state, threatened by 'herd mentality' and 'media-star' scientists. Locke felt so strongly about this that he joined forces with Peter Lawrence from the MRC Laboratory of Molecular Biology in Cambridge to publish a commentary to this effect in *Nature*.[4] And Locke and Lawrence are not the only modern scientists to laud Wigglesworth for his dedication, humility, and extraordinary contribution to learning. In 2002, some eight years after Wigglesworth's death, Serap Akman of Yale University School of Medicine and his colleagues named a genus of bacterium that lives in the gut of the blood-sucking tsetse fly in his honour.[5]

It is unlikely that any such tribute will be afforded Donald Williamson in his lifetime. As we have seen, science is conservative by nature and it is almost inevitable that the majority of scientists will mistrust iconoclasm. But iconoclasm is also important for the advancement of science: without it we would still believe the Sun orbited the Earth and that the continents have always occupied the positions they do today. Iconoclasm, in its turn, operates by challenging orthodoxy with novel hypotheses and theories. In the introduction to his seminal book on the structure of scientific revolution, Thomas S. Kuhn explains how people have been misled by an overly simplistic understanding of how scientific theories develop. Most of us, whether scientists or non-scientists, come across theories through studying textbooks, attending lectures, tuning in to radio or television programmes, or reading articles and communications long after the theory was first conceived and has been established as "fact". In Kuhn's words, 'the aim of such [communication] ... is persuasive and pedagogic; a concept of science drawn from them is no more likely to fit the enterprise that produced them than an image of a national culture drawn from a tourist brochure'.[6] Williamson's story, whether one agrees with his theory or not, is illuminative of the concept of "ideational evolution".

We have glimpsed, at various stages during its development, how Williamson's initial idea of larval transfer changed and expanded, evolving, through the confirmation of experiment, from hypothesis to theory. 'Progress in knowledge', as the late renowned palaeontologist Stephen Jay Gould once explained, 'is not a tower to heaven built of bricks from the bottom up, but a product of impasse and breakthrough, yielding a bizarre and circuitous structure that ultimately rises nonetheless'.[7]

In mid-January 2002, as Williamson led me on an inspection tour of the tank-holding facilities at the Port Erin Marine Laboratory, we were discussing the history of his theory. We walked across wooden ramps between a variety of holding facilities in the old and somewhat dilapidated indoor labs – already heralding a still gloomier fate – where I marvelled at spider crabs looking back at me from their perches on stones; scallops within their shells, an almost perfect encapsulation of the logo of the multinational oil company; starfish; sea urchins; octopuses. In my ears were the rushings of water, while my nostrils were overwhelmed with the brine of the oceans.

By that time, I had come to agree with Williamson that there seemed to be many anomalies in marine metamorphosis, anomalies that were difficult to explain along orthodox evolutionary lines. I knew this perception had led him to construct a novel if highly controversial explanation, one that was continuing to evolve even then. His theory, in so far as I was aware, was the only one that attempted a universal explanation of metamorphosis throughout the animal kingdom. This in itself seemed worthy of exploration, and so I asked him the inevitable question.

'You wrote the theory in a series of papers and then in book form, in 1992. But even after that, you subsequently expanded the theory in a really big way. I just wondered why you felt you needed to do that.'

'Well, there was no dramatic breakthrough. It was very gradual. I realised when I wrote the original paper, and the book, that this was not the end. There must be lots more cases that I had not yet considered. I started to consider the crustaceans in greater detail. I was already a crustacean specialist. I decided that most crustacean larvae had been transferred from other groups. I was still, as I then saw it, hanging on to the probability that creatures that metamorphosed in a more gradual way were part of the same genome as the adult, whereas those that showed a drastic metamorphosis certainly were not. A crab hatches out as a so-called zoea larva, which is nothing whatever like a crab. On the other hand, a megalopa larva is quite like a crab.'[8] Here, he referred to what is often, but not always, the final larval stage in crustaceans, a larva with the many appendages and claws in place and often a single tail for locomotion. The zoea, in contrast, may be slightly reminiscent of the horseshoe crab (an unrelated arthropod) or a shrimp.

'Yet the metamorphosis of zoea to juvenile crab is very gradual', he continued. 'At that stage I thought the megalopa was part of the original genome and the zoea was from another source entirely. I would now say that the megalopa is also from another source, another vaguely crab-like animal that doesn't need a big metamorphosis. I then wrote a paper on the evolution of crustacean larvae. I could not draw for myself so I got Tony Rice to provide the illustrations. I sent it in to *Crustaceana*, the specialised journal, and I was very surprised when they accepted it, almost immediately and without a quibble. Anyway, from that time on I was extending it gradually with the idea that nearly all larvae had been acquired from other sources, but I was hanging on to the possibility that it wasn't quite all.'[9]

In the laboratory I found myself gazing at *Alcyonium digitatum*, known as dead man's fingers, shaped like a stumpy, white, bloodless hand adorned with numerous sausage-shaped fingers. These are coral-like animals that, when feeding, produce numerous tiny polyps, giving their surface a fuzzy appearance thought to resemble decaying flesh. Nearby, I almost crunched my feet on some "mermaid's purses", quadrangular shapes about five centimetres long, brown and gelatinous, that encase the eggs of dogfish and many species of sharks.

Our conversation continued as we left the indoor lab and walked round the back of the building to where some large out-door tanks were set down in the shadows of rocky cliffs. Seagulls wheeled and cried above our heads. Williamson pointed to a bricked tank, built into the cliff. This tank had hosted his *Ascidia mentula* up to 1990, but it was leaking even then and soon after-wards it was judged beyond repair. I found myself gazing up at the huge concrete-walled tank that had replaced it, currently being used to spawn fish. From here the larval forms or small fish were being delivered to people who needed them to restock marine areas. Not long after I was there, this same tank would provide the *Ciona intestinalis* and *Ascidiella aspersa* for the 2002 experiments; for further studies of *Ascidia mentula*, divers would have to journey out and explore a shipwreck three miles north-west of Bradda Head.

At the time of my visit, Williamson was working towards the publication of a second book, *The Origins of Larvae*, in which he would extrapolate his theory of larval transfer to embrace all forms of marine invertebrate larvae and, in subsequent years, in what would prove to be even more controversial, he would further extend it to include metamorphosis in insects.[10] In his early work, including his first book, *Larvae and Evolution*, he had confined himself to his chosen discipline of marine biol-ogy, however controversial his ideas might have seemed to colleagues. He had put forward a consistent hypothesis – for example, about the radical shifts in axes of development and symmetry, from bilateral to radial – and he had devised and conducted experiments aimed at confronting that hypothesis.

In comparison, this later expansion seemed, to me, less consistently logical, and with the inclusion of insects it pulled him beyond his own particular scientific expertise. Unlike his earlier theory, he had performed no probing experiments involving insects.

'I'm still not quite clear in my mind why you thought that was the case,' I said to him, questioning the ambition of the newer book.

'Well, I was reluctant at that time to go the whole way. When I wrote the original paper I was more or less convinced that [larval transfer] applied only to the exceptional cases. Other larvae had evolved from the same source. The crustacean paper was a big breakthrough. This was followed by the Fallen Leaf Lake Conference.'

In September 1996, roughly four years after publication of *Larvae and Evolution*, Williamson had been contacted by Michael Syvanen, Professor of Medical Microbiology and Immunology at the School of Medicine of the University of California, Davis, to contribute to a conference on horizontal gene transfer at Fallen Leaf, California.

'I don't know how I got into contact with Mike Syvanen. He's the "horizontal gene transfer man". Lots of work had been done on bacteria and there was no doubt that genes could be transferred between bacteria; he was delighted to get an example of a theory of gene transfer in eukaryotic animals. So he invited me to his conference in California.

'Enid and I hemmed and hawed about whether we could get there. Eventually we decided that we couldn't. So I went to Liverpool, to the university, and I got them to make a video of the theory as it then was. The video did not go much further than my 1992 book. I sent [Syvanen] a copy of this, which went down very well ... Then I was invited to write an article on the same theory for the book that followed [the conference]. I started off more or less writing what was in the video.

'In the meantime, I had written the crustacean paper. This was already in press. I was now happy with the idea that larval transfer was much wider in application than in the original paper or the book. Each time I sent Syvanen my manuscript, I would follow this with a letter to say that a revised version would follow. Each version traced another larval form to an adult in another taxon: trochophores to rotifers resembling *Trochosphaera*, and tornarias to an ancestor of *Planctosphaera*. This meant that Johannes Müller's larva is not the primal trochophore, and acorn worms did not invent their own larvae. All larvae originated as adults in foreign taxa. The horizontal gene transfer book was published in 1998, with a revised edition in 2002.[11]

'By that time, I had decided that all drastic metamorphoses were due to the larval form being acquired from another group. But I was still undecided about the more gradual metamorphoses. It was while I was writing the chapter in Syvanen's book that I got the idea that this could be extended further, not only in the crustaceans but also to different groups.'

'You kept extending the application of your theory?'

'Yes. I kept extending it. I eventually reached the stage where I felt I could extend it not only to all larvae but to all embryos. Again, it just seemed to fit, so why not? There is no other explanation for where embryos come from, so why not larval transfer, or some similar horizontal genetic transfer?'

We have seen that at the heart of biological theory there is a great divide on the origins of metamorphosis. On the one hand, we have the "Haeckelian theory", in modern form best exemplified by the work of Swedish biologists Gösta Jägersten and Claus Nielsen, which takes the view that metamorphosis is ancestral to the evolution of all animal life. On the other hand, we have the "direct development theory", proposed by Williamson, in which the earliest animals did not metamorphose.

In the Hackelian theory, the larval to adult cycles evolved at the very beginnings of animal life and led, within the same individual life cycle, to a free-swimming "pelagic" phase, spent in the layers of the ocean closer to the surface, that would become the primary larva, and to a bottom-dwelling "benthic" phase that would become the adult. This theory suggests that there were survival advantages in a dual life cycle, with the pelagic forms better equipped to disperse across wider distances and into new habitats in the three-dimensional ecosphere of moving oceans, while the benthic forms were better equipped to explore the rich food sources and other potentials of the two-dimensional ecosphere of the ocean floor. This view rests upon two important assumptions. It takes for granted that primary larvae evolved through a process of linear descent with modification from the earliest and simplest of animal forms, dating back to the Precambrian period. It also implies that all the complex life cycles that we see today must have come about as a result of linear descent with modification of that primal cycle. But is this a reasonable working theory?

In my opinion, this theory can only founder if it assumes that the exclusive mechanism of evolutionary change is that of mutation under the direction of natural selection, which has very often been the assumption of the past. If, however, we enlarge the theory to embrace all of the mechanisms for hereditary change, including mutation, hybridisation, genetic symbiosis, and epigenetics – and if we include the operation of such mechanisms during embryonic development – with natural selection working at different levels, then it becomes a good deal more reasonable. Still, such a theory would be unlikely to unite the world of evolutionary biology. Based on the analyses of Richard Strathmann, and the continuing debate at large in the field of developmental biology, it would likely struggle when confronted with the vast range of metamorphosis. For instance, an original metamorphic life cycle could not explain the origins of complete metamorphosis in insects, since most entomologists are agreed that metamorphosis is not ancestral to insects; it evolved long after the ancestral forms of the first insects invaded dry land. Moreover, a significant number of marine biologists think that many of the larval forms seen today in marine invertebrates evolved after their adult forms did.

If James Truman and Lynn Riddiford are right, the break-through to the dramatic change of complete metamorphosis in insects may have come about when one group evolved the abil-ity to hatch out early, releasing an offspring at a premature but nevertheless viable stage in its embryonic development. What, in such a scenario, might have given rise to an inherited pat-tern of viable premature hatching? One possibility is a muta-tion, or possibly a series of mutations, affecting developmen-tal pathways. But it is important that we continue to consider the other mechanisms itemised above: not merely mutation, but also hybridogenesis, symbiogenesis, and epigenetics, even if we accept the basic hypothesis presented by the scientific partners.

Williamson refutes all such extrapolations of the Haeckelian theory. He disagrees with the theory that larvae originated in a linear descent-with-modification fashion as part of a single life cycle spent in two phases, the first spent swimming around the primal ocean currents and the other spent lurking for food on the ocean floor. Instead he proposes that some animals evolved an entirely pelagic life cycle while others evolved entirely ben-thic cycles. In his view, marine metamorphosis came about not within a single animal's life cycle but through hybridisation between these two dissimilar life cycles, resulting in a sudden saltation and a single descendant organism.

It is important to realise that, in Williamson's opinion, this did not happen in the recent past. On the contrary, he proposes that such hybridisations happened long ago, when genomes might have been less honed by selection. The frequency and diversity of hybridisations would undoubtedly have been facilitated by the broadcast spawning so prevalent in marine life forms. How varied must have been the outcomes in terms of new animal forms and lifetimes through millions of putative hybridisations over vast periods of time, with long-term success or failure rewarding those with the greatest selective advantage not only in terms of form but also in terms of life cycle. In his attempts at cross-phyletic hybridisation, he realised it was perhaps too much to hope that over a small number of experiments he could recapture an occasional event from the very distant past. All he really dared to hope was that that some form of hybridisation would occur – and be recordable.

We need to ask ourselves not merely if Williamson's theory of metamorphosis by hybridisation is true, but whether his theory and work contribute, specifically or generally, to scientific understanding.

We might recall Sebastian Holmes's response when I asked him what was needed to confer plausibility to Don Williamson's theory. Holmes felt Williamson had already done enough to establish this, but, together with Nic Boerboom, Holmes had combined forces with Williamson to replicate the earlier experiments, producing a previously unknown, asexually reproducing life form – the budding spheroid. This result has not, to date, been confirmed by a genetic analysis of the hybrid offspring. Further study is surely warranted. Moreover, none would disagree with Richard Strathmann that such further study must include molecular and genetic analysis – not a superficial examination of one or two genes, but a rigorous and meticulous analysis, stage by stage, at the level of painstaking detail pioneered by Rudolph A. Raff at Indiana University.

Such genetic and molecular studies should readily confirm or refute that the spheroids are genuine cross-phyletic hybrids.

And if confirmed, much might be learnt of unknown systems of genetic and epigenetic control. Indeed we simply do not know what to expect – and that in itself might make the cost and effort of such a major undertaking eminently worthwhile.

There is an additional pressing reason why we should consider Williamson's broader viewpoint. In the words of Stephen Jay Gould, 'The pleasure of discovery in science derives not only from the satisfaction of new explanations, but also, if not more so, in fresh (and often more difficult) puzzles that the novel solutions generate'.[12] And it is in consideration of Gould's premise that I shall close this account of Williamson's strange, exciting, and always probing exploration, this time with his expansion of hybridisation and its role in metamorphosis to help explain a second mystery, one less familiar to lay readers than that of metamorphosis, yet one that has loomed over the pursuit of zoology for a century or more. This is the mystery of the origins of the entire animal kingdom, known to biologists as the Cambrian explosion.

The Burgess Shale is a fossil bed in the Canadian Rocky Mountains, dating to the Middle Cambrian period, some 545 to 525 million years ago. In this fossil bed, and in an increasing number of related deposits elsewhere throughout the world, the forerunners of modern animals appear for the first time. As has become increasingly clear, their arrival in the fossil record was not heralded by a vast period of progressive evolution among the various animal groups. Instead there seems to have been an amazingly rapid and extraordinary flowering of different branches of the tree of life, during which, as Gould expresses it in his book, *Wonderful Life*, 'modern multicellular animals make their first uncontested appearance – and with a bang, not a protracted crescendo'.[13]

For the first three and a half billion years that life existed on Earth, it remained in the very basic form of bacteria. By two and a half billion years ago, bacteria had evolved the ability to capture the energy of sunlight through photosynthesis, ushering in the Proterozoic era, which would extend to the appearance of the first multicellular animal forms during the Vendian period, from 600 million to 545 million years ago.[14] So-called Vendian forms originally referred to Precambrian fossils found in the White Sea area of Russia, just north of Archangel, now known as Arkhangel, while another group of early animal forms, called the Ediacaran fauna, referred to slightly earlier Precambrian fossils found in the Ediacaran Hills of South Australia. There is a confusing tendency to mix terms and dates, but even if we broadly attribute the start of the Precambrian to about 600 million years ago and of the Cambrian to about 545 million years ago, the intervening period, some 55 million years, is, in geological time, a brief interval in which to imagine the first appearance of virtually all of the animal phyla.

Some of the earliest Precambrian fossils are little more than blobs. Others appear to represent early cnidarians, like the present-day jellyfish, sea pens, and sea anemones. Among the many enigmatic fossils discovered in Ediacaran rocks is an impression of the animal *Arkarua adami*, named for Arkarua, the mythical giant snake of the Aboriginal peoples who live close to where the fossil was discovered, in the Flinders Ranges of South Australia.[15] *Arkarua* is a small disc-like fossil with a body that is divided into five lobes – pentaradial. It may represent the earliest echinoderm, which, if true, would confirm Williamson's contention that the first echinoderms were radial.[16] Unfortunately, the fossil lacks soft tissue details, so the link to the echinoderms is only speculative at present.

In a much broader respect, the Cambrian explosion gives added credibility to Williamson's emphasis on the role of hybridisation in marine metamorphosis and animal evolution more generally. Suddenly, from a geological perspective – Simon Conway Morris, Professor of Evolutionary Palaeobiology at the University of Cambridge, brackets the principal events of evolutionary interest as taking place between 550 million and 530 million years ago – a massive diversity of animal forms appear in the fossil record.[17] The sudden appearance is truly dramatic. Other scientists, reporting on the fossil record of a contemporaneous site in Yunnan Province, China, that is known as the Chengjiang biota, have looked specifically at the development of animals with exoskeletons and the chordate beginnings of endoskeletons, and have proposed that most of these forms initially arose over a period of a just ten million years.[18]

Any universal theory of evolution is obliged to explain how, over such a short period of time, the body forms of most present-day, and many extinct, animal phyla came into being. But no reasonable thesis has ever been put forward. On the contrary, the Cambrian explosion has been a source of controversy since it was first observed by geologists in the 1830s. Charles Darwin was well aware of its challenges to his ideas of a gradualist evolution, and he devoted the ninth chapter of *Origin of Species* to what he termed 'the imperfection of the geological record', concluding that if his theory were true, it demanded that a period of gradual evolution must have preceded the Cambrian explosion, '… as long as, or probably far longer than, the whole interval from the Siluran [Cambrian] age to the present day; and that during these vast, yet quite unknown periods of time, the world swarmed with living creatures'. He added, 'To the question why we do not find records of these vast primordial periods, I can give no satisfactory answer'.[19] Today, there is no escaping the fact that Darwin was mistaken. The apparent explosion is not the result of imperfections of the fossil record; the much better fossil record we have found since then confirms that the explosion is real.

The animals fossilised in the Burgess Shale do not date to

the Cambrian explosion itself, but they appeared soon after it, '... before the relentless motor of extinction had done much work', as Gould describes it. This gives the fossils, and similar groups discovered worldwide, including at Chengjiang, iconic importance. Moreover, their preservation is all the more precious in revealing not only fossilised skeletons but also exquisite detail of the animals' soft-tissue forms. While many of these forms are ancestral to modern animal phyla, a number of them look bizarre – and do not seem to have a readily discernable legacy among living animals. Thus, these Vendian and Cambrian fossils offer an extraordinary snapshot of animal evolution during and very soon after its early diversification. Perhaps we should consider what we are privileged to witness here: not circus monsters with names such as *Hallucigenia* (for "dreamlike") and *Anomalocaris* (for "abnormal shrimp"), but true wonders of life at its most creative. These forms suggest intriguing evolutionary experiment, arising from what Gould describes as a 'staggeringly improbable series of events ... utterly unpredictable and quite unrepeatable'.[20]

One such creature is *Pikaia gracilens*, a slender lanceolate creature about four centimetres in length, with a fish-like musculature, a notochord, and a peculiarly shaped head, which is cleaved into two fleshy extensions, or lobes. Simon Conway Morris believes this animal, like the slightly older *Cathaymyrus diadexus* found in the Chengjiang biota, must be one of the earliest chordates.[21] *Pikaia* likely swam through the water, driven by undulations of its body, much like a fish or an eel; the lobes may have been used to filter food.

The most common fossil animal found in the Burgess Shale – and the first collected by the eminent American palaeontologist Charles Doolittle Walcott when he discovered the field in 1909 – is *Marrella splendens*, with more than fifteen thousand examples catalogued to date. A tiny arthropod no bigger than a baby shrimp, it is striking, with a head shield decorated with two broad, horseshoe-shaped antennae. Its body is composed of between twenty-four and twenty-six segments, each sprouting pairs of branched appendages, the upper ones a long feathery gill for breathing and the lower ones twenty-four to twenty-six pairs of legs. Its mouth parts, unfortunately, remain unknown. When Walcott found it, he thought it represented an extinct species of trilobite, one of the most common animals seen in the fossil record. But Harry B. Whittington, Professor of Geology at the University of Cambridge, re-appraised it in 1971 and concluded that it just did not fit into any of the known modern phyla.[22] Today, *Marrella* is considered to represent a distinct order within the superphylum of the arthropods.

Another markedly beautiful animal, *Burgessia bella*, is named for the fossil site itself, and is the third most common arthropod found there. When viewed from above, *Burgessia* appears to be encased in an oval carapace, somewhat resembling the head shield of a horseshoe crab, though only a fraction of its size; two delicately curled antennae project from the front and an elongated spike of a tail, twice the length of the carapace, projects from the rear. Underneath the carapace, the body is divided into two distinct sections. At the front, immediately behind the sprouting antennae, is a mouth devoid of jaws, and, to date, palaeontologists have found no definite evidence of eyes. The front section also bears three sets of legs. In the rear, there are seven branched appendages, each dividing into outer gills and ground-facing legs. C.P. Hughes, who re-examined the fossilised animal in 1975, wrote, 'What is apparent from this restudy is that *Burgessia* possess a mixture of characters ... many of which are to be found in modern arthropods of various groups'.[23] Gould remarks that Hughes felt unable to place *Burgessia* in the standard taxonomic tree, 'because he regarded this genus as a peculiar grab-bag, combining features generally regarded as belonging to a number of separate taxonomic groups'.[24]

In his charming book, *The Crucible of Creation*, Simon Conway Morris imagined two time-travelling zoologists as they gaze upon these extraordinary creatures, alive and flourishing in the shallow oceans. He recreates an animal, *Nectocaris pteryx*, in exquisite detail, having once himself dissected its single known fossil from the Burgess rock.[25] What Morris glimpses in the fossil come to life is a very peculiar creature indeed, with enormous eyes set right in front of a carapace-shrouded head. But if the head is reminiscent of that of an arthropod, the abdomen, with its prominent longitudinal fins and fin rays, appears to belong to an entirely different phylum. In his words, 'Such a feature is never seen in the arthropods'.[26] It is, however, seen in the modern oceans: arrow worms are speedy and efficient hunters, with streamlined bodies and fins supported by fin rays, similar to those of *Nectocaris*. But otherwise, *Nectocaris* bears little similarity to the arrow worm. The simple truth is that its front resembles an arthropod while its back resembles a chordate, complete with tail and fin. 'What can be done with such a chimera!' Gould remarked.[27]

Again and again, in the reappraisals of animal fossils from the Cambrian explosion, we encounter what Williamson would describe as significant anomalies. According to Gould 'Thus, the bivalved arthropods – the group that seemed most promising as a coherent set of evolutionary cousins – also formed an artificial category hiding an unanticipated anatomical disparity'. What possible order could be found in the Burgess arthropods? 'Each one seemed to be built from a grab-bag of characters – as though the Burgess architect owned a sack of all possible arthropod structures, and reached in at random to pick one variation ... whenever he wanted to build a new creature'.[28] In the opinion of James W. Valentine, Professor Emeritus at the University of California, Berkeley, and a distinguished evolutionary palaeontologist, the explosion in species must reflect 'some very special circumstances in life's history'.[29]

It is hardly surprising that this explosion has provoked
a remarkable range of speculative explanations – indeed so
remarkable and varied that when Valentine tried to organise
all such hypotheses into a comprehensive framework, he was
unable to do so. Instead, he rounded it all off as best he could
to a small number of representative ideas. Some biologists
merely dismissed the reality of the Cambrian explosion; others
attributed it to physical changes in the environment, including rapid rise of atmospheric oxygen; others built hypotheses
based on biological changes in the environment, such as the
consequences of the rise of new planktic forms and the evolution of predators; and others still put it down to what Valentine
labelled "intrinsic evolutionary change". Surely common sense
would suggest that, whatever the contribution of the physical
environment or the biological component of the ecology, any
reasonable explanation of the Cambrian explosion must come
from the known mechanisms of evolutionary biology.

In *Origins of Larvae*, Williamson highlights another anomaly,
this time not larval but appearing in the body plan of several
groups of animals, known as lophophorates, for the circular or
horseshoe-shaped ring of tentacles, the lophophore, they use to
trap their food. The lophophorates include four marine invertebrate phyla: the bryozoan, or moss, animals, which are tiny
filter feeders; the phoronids, which are bottom-feeding worms;
the brachiopods, also known as lampshells because they resemble ancient oil lamps in having their bodies hidden between
two unequally sized clam-shaped shells; and finally, the ento-
procts, tiny transparent marine animals previously classed with
the bryozoans but excluded these days because the anus as well
as the mouth opens within the lophophore.

Within the phylum of the hemichordates – we might recall
that they are a group of headless chordates that includes the
acorn worms – the pterobranchs also feed using a lophophore.
In Williamson's view, the lophophore is such a complex organ

it is unlikely to have evolved more than once. And this, in turn, would suggest that the various, seemingly unrelated groups of animals that possess a lophophore must in fact be related to one another through some common evolutionary ancestor. Pat Willmer, the distinguished biologist at the University of St Andrews, acknowledges that the lophophorates seem to belong together, and to stand somewhat apart from the other members of the phyla. But the closer she looks, the more confusing the picture becomes. 'Most importantly, they show a very complex mixture of the set of features that are normally used to define protostomes and deuterostomes, thus giving them an apparently anomalous position somewhere in the middle of the animal kingdom'. She goes so far as to conjecture that the lophophorates might be 'the transitional stage in the "invention" of the deuterostomes'.[30]

In ruminating over this quandary, Williamson wondered if these difficulties could be reconciled if the lophophorate developmental plan had been transferred between taxa in the distant past through hybridisation. If that were the case, it would open the floodgates to many other possibilities. What if there had been sporadic hybridisations between organisms ever since the invention of sex – an event that long preceded multicellular animals? Sexual crossing between disparate species of animals at a very early stage in their evolution might have resulted in the blending of body forms and components. The resulting offspring, and their descendants, would truly be chimeras – amalgamations of quite different species. 'The more I thought of it, the more it made sense'.

In April 2004, at the invitation of Lynn Margulis and Wolfgang Krumbein, one of the founders of the field of geomicrobiology, Williamson spoke about these new ideas at a scientific conference in Bellagio, Italy.[31] Here he talked about the expansion of his hybridisation theory to lophophorates for the first time. In December of the same year, Margulis suggested that Williamson should contribute to an on-line cultural forum, *The Edge*, aimed at broadcasting cusp-of-the-wave debate and enlightenment; each New Year, the forum set out a broad but seminal question to which various thinkers might freely contribute a personal answer. *The Edge* question for the opening of 2005 was: *What I believe but cannot prove*.[32] In Williamson's words, 'I said, in half a page, that I believed I could explain the Cambrian explosion'.

Valentine, in appraising the first appearance of what we now recognise as phyla in the early Cambrian fossil record, was convinced that the fossil record really was consistent with the actual evolutionary events. In his book *On the Origin of Phyla*, he states: 'The abrupt appearance of disparate body types without a record of intermediate forms ... suggests that significant morphological change may evolve very rapidly'.[33] In case his readers should harbour any lingering doubts, he emphasises what this means: 'The implication is that the abrupt appearance of many phyla during the [early Cambrian] could mirror an actual rapid radiation of body plans, just before or during that time period'. Valentine's view is widely shared among evolutionary biologists. So how could such major body changes evolve in such a brief time period? While we can reasonably assume that all of the mechanisms of genomic creativity have played some part, one such mechanism, hybridisation, is an essential source of rapid, large-scale change. And now this force, so long neglected in the study of animal evolution, sprang to Williamson's mind.

Williamson, like most biologists, had long been puzzled by the mystery of the Cambrian explosion. 'Not until the summer of 2004 did I read Stephen Jay Gould's *Wonderful Life* on the Burgess Shale fauna. Now I decided that these remarkable animals were probably the results of component transfer – they were concurrent chimeras'. Williamson wrote a new article, with the provocative title, "Hybridisation in the Origin of Animal Form and Life-cycles".[34] In this paper, published in 2006, he extended his hybridisation theory to embrace both larval transfer and the Cambrian explosion, developing his ideas along lines that will now be familiar. The early animals were exclusively marine. They shed their eggs and sperm into the water, where fertilisation took place. At such a time, when genomes were presumably more malleable to hybrid union than today, hybrid unions must have been common, not only between the first animals but also between the resulting hybrids. 'This produced many concurrent chimeras, which included the first members of most animal phyla, living and extinct'.

Williamson acknowledges the significance of additional factors that might have encouraged genomic experimentation at the time of the Cambrian explosion. Palaeontologists believe the Cambrian explosion followed the great Precambrian extinction, dated at about 650 million years ago, possibly the greatest devastation of life on Earth. The same event may have led to the catastrophe known popularly as "Snowball Earth", when approximately seventy per cent of the Precambrian flora and fauna appear to have perished. Some biologists believe this was followed by a lesser extinction, at the end of the Vendian period, that may have wiped out a great many of the remaining survivor species. Whatever the cause, or causes, of these catastrophes, many ecological opportunities would have arrived in the wake of these extinctions, opportunities that would have fostered the survival of the newly evolving animals of the early Cambrian. Some authorities, such as Valentine and Morris, have suggested there was an exceptional availability of food and nutritional sources, which contributed to the proliferation of life.

There is some evidence that the oceans, in early Cambrian times, contained relatively high levels of phosphorus, a compound that modern-day oceanic and terrestrial farmers would be altogether familiar with as a prerequisite for rapid and healthy biological growth.[35] Natural selection will have determined the fate of the many contemporaneous genomic experiments, leading to the animal phyla we are familiar with today. Others that subsequently became extinct survived long enough to be captured, in spectacular and beautiful detail, in the fossil record.

In 2010, proposing that the same force, hybridisation, was capable of explaining both the origins of larvae and the origins of the animal phyla during the Cambrian explosion, Williamson brought together the two inter-related strands of his theory in a new publication.[36] This extension of his hybridisation theory to the origins of the animal kingdom soon proved every bit as controversial as his larval transfer theory. And exactly the same questions apply, in biological and evolutionary terms. Does the theory have sufficient validity to make further testing reasonable and worthwhile?

With the availability of whole genome sequences, Michael Syvanen, the horizontal gene transfer expert, has begun a systematic evaluation of the evolutionary origins of the major phyla, including both plants and animals, re-appraising the various basic groups and their places on the tree of life. In a paper published in 1985, when Syvanen was working at the Harvard Medical School, he proposed that horizontal transfer of genes between different taxa across the evolutionary tree – already observed to take place between different groups of bacteria – was likely to apply to all life. Such transfers would bring pre-evolved genes from different evolutionary lineages together in sudden evolutionary bursts, and this might help to explain, at least in part, the discovery of similar genes throughout all life, as well as rapid bursts of evolution, such as the Cambrian explosion. In 2010, Syvanen and Jonathan Ducore, now working in the Departments of Microbiology and Pediatrics at the University of California, Davis, School of Medicine, compared and contrasted the whole genomic sequences from some of the major animal phyla. Their conclusions challenged some of the prevailing assumptions.[37]

In essence, evolutionary theory predicted that phyla with closer evolutionary links to one another would inevitably have more genes in common. These similarities should appear in proteins, since protein sequences derive directly from genetic sequences. Drawing upon the expanding libraries of whole genomes, Syvanen and Ducore examined the sequences of two thousand different proteins. Next, counting the numbers of proteins in each separate phylum, they grouped the animal phyla into evolutionarily-linked clusters – clades – based on the numbers of proteins they shared. Extrapolating from the conventional tree of life, such an exercise should confirm the evolutionary commonalities of the deuterostomes, including the sea squirts with their tadpole larvae, the sea urchins with their pluteus larvae, and humans. But when the researchers looked for evolutionary commonalities among the two thousand proteins available to them, they saw that the protein sequences fell into two distinct groups. About half supported the notion that the urochordates were indeed related to the chordates; the other half suggested that the urochordates descended from an ancestor whose closest living relatives are found amoung the arthropods and nematodes, that is, with the protostomes – in other words, a group that had no common ancestry with the chordates.

These results suggested that a major horizontal gene-transfer event had taken place during the emergence of one of the animal phyla. The simplest explanation, in the opinions of the authors, was that the ascidians began as a hybrid between a

primitive chordate and some other organism, perhaps 'from an
extinct and unidentified protostome phylum, at a time close to
but after the diversification of the chordates and echinoderms'.
As they further noted: 'We are not the first to be intrigued with
tunicate taxonomy. The tunicate anomaly has perplexed students of biology for more than a century and led Don Williamson to suggest that the tunicate had hybrid origins with those
genes controlling larval development coming from a chordate
ancestor, and those genes controlling adult development coming from some other phylum'.

Syvanen and Ducore's research is pioneering in its thinking and methodology, and their findings, too, are likely to be
seen as controversial. Other scientists are poised to take up
their innovative methodology to confront their findings – as
indeed they should. But if, in time, Syvanen and Ducore's conclusions are confirmed, it will demonstrate the importance of
hybridogenesis as a mechanism for major, and relatively rapid,
transformations at the very root of animal evolution. It will also
illustrate the importance of iconoclastic thinking in science, the
sort of thinking that has characterised Williamson's approach.

This, however, does not mean that evolutionary science
has been misguided thus far. Mutation, in the original sense
of genetic errors that arise in the copying of DNA during cell
division, continues to be vital in understanding the genetics of
hereditary change, and it continues to offer clues to the genetic
underpinnings of many diseases. But mutation is insufficient on
its own to explain the full range of hereditary change. Hybri-
dologists don't regard hybridogenesis as a mutation, any more
than symbiologists regard genetic symbiosis, or epigeneticists
regard epigenetic change, as such. Modern evolutionary biology, including the rising field of evo-devo, must surely embrace
all of these mechanisms. Through the work of a new generation
of scientists, who are aided by increasing knowledge of whole
genomes, evolutionary science is coming to grips with some of

the great mysteries of biology, including metamorphosis and the Cambrian explosion.

I don't believe science, through investigation and explanation, removes the sense of awe and wonder that comes from beholding a great natural mystery. On the contrary, it deepens it. How extraordinary if, through the trials and errors of evolution, over immense periods of time, the natural world has performed quasi-miracles of metamorphosis every bit as imaginative and magical as those dreamt up in Ovid's poetic imagination.

# Bibliography and Further Reading

Alcock, John. *In a Desert Garden: Love and Death among the Insects*. Tucson: University of Arizona Press, 1999.

Arnold, M.L. *Natural Hybridisation and Evolution*. Oxford: Oxford University Press, 1997.

Balfour, Francis M. *A Treatise on Comparative Embryology*. vol 2. London: Macmillan & Co, 1880–1.

Bonner, John Tyler. *Lives of a Biologist: Adventures in a Century of Extraordinary Science*. Cambridge, Mass.: Harvard University Press, 2002.

Boucher, Douglas H., ed. *The Biology of Mutualism: Ecology and Evolution*. Oxford and New York: Oxford University Press, 1985.

Bronowski, Jacob. *The Ascent of Man*. BBC and Time-Life Films, 1973.

Clark, Catherine M., and James Macalister Mackintosh. *The School and the Site: A Historical Memoir to Celebrate the Twenty-fifth Anniversary of the School*. London: H.K Lewis & Co. Ltd., 1954.

Daniel, R.J., ed. *Lancashire Sea-Fisheries Laboratory: James Johnstone Memorial Volume*. Liverpool: Liverpool University Press, 1934.

Darwin, Charles. *On the Origin of Species by Means of Natural Selection* (reprint). London: Penguin, 1982.

———. *The Descent of Man* (reprint). New York: Prometheus Books, 1998.

Davidson, Eric H. *Genomic Regulatory Systems: Development and Evolution*. San Diego: Academic Press, 2001.

Dawkins, Richard. *The Selfish Gene* (reprint). Oxford and New York: Oxford University Press, 2006.

Etkin, William, and Lawrence I. Gilbert, eds. *Metamorphosis: a Problem in Developmental Biology*. New York: Appleton-Century-Crofts, 1968.

Fabre, Jean-Henri. 'The Life of the Caterpillar.' In *Souvenirs Entomologiques*. Trans. by Alexander Teixeira de Mattos, London: Hodder and Stoughton, 1912.

Fisher, Ronald Aylmer. *The Genetical Theory of Natural Selection*. Oxford: Clarendon Press, 1930.

Garstang, Walter. *Larval Forms and Other Zoological Verses* (reprint). Foreword by Michael LaBarbera. Chicago: University of Chicago Press, 1985.

Gehring, Walter J. *Master Control Genes in Development and Evolution: The Homeobox Story.* New Haven, Ct.: Yale University Press, 1998.

Gilbert, Lawrence I., et al., eds. *Metamorphosis: Postembryonic Reprogramming of Gene Expression in Amphibian and Insect Cells.* San Diego: Academic Press, 1996.

Gilbert, Scott F. *Developmental Biology.* Seventh Edition Sunderland, Mass.: Sinauer Associates, 2003.

Gilbert, Scott F., and David Epel. *Ecological Developmental Biology: Integrating Epigenetics, Medicine, and Evolution.* Sunderland, Mass.: Sinauer Associates, 2009.

Giudice, Giovanni. *Developmental Biology of the Sea Urchin Embryo.* New York: Academic Press, 1973.

Gould, Stephen Jay. *Wonderful Life: The Burgess Shale and the Nature of History* (paperback reprint). London: Penguin, 1991.

——. *Eight Little Piggies: Reflections in Natural History.* London: Penguin, 1994.

Haldane, J.B.S. *The Causes of Evolution.* London and New York: Longmans, Green & Co., 1932.

Hall, Brian K., and Marvalee H. Wake, eds. *The Origin and Evolution of Larval Forms.* San Diego: Academic Press, 1999.

Hoshi, Motonori and Okitsugu Yamashita, eds. *Advances in Invertebrate Reproduction, 5: Fifth International Congress of Invertebrate Reproduction, Nagoya, Japan, July 23–28, 1989.* Amsterdam and New York: Elsevier, 1990.

Huxley, Julian. *Evolution: The Modern Synthesis.* London: G. Allen & Unwin Ltd., 1942.

Huxley, Thomas Henry. "The Darwinian Hypothesis." In *Darwiniana: Collected Essays* vol. 2. London: Macmillan, 1893.

Hyman, Libbie Henrietta. *The Invertebrates* vol. 1–5. New York: McGraw-Hill, 1940–9.

Jägersten, Gösta. *Evolution of the Metazoan Life Cycle: A Comprehensive Theory.* London and New York: Academic Press, 1972.

Judson, Horace F. *The Eighth Day of Creation: Makers of the Revolution in Biology* (paperback reprint). London: Penguin, 1995.

Kuhn, Thomas S. *The Structure of Scientific Revolutions,* 3rd ed. Chicago: University of Chicago Press, 1996.

Lewis, Walter H., ed. *Polyploidy: Biological Relevance.* New York: Plenum Press, 1980.

Lotsy, Johannes Paulus. *Evolution By Means of Hybridisation.* The Hague: M. Nijhoff, 1916.

Margulis, Lynn. *Origin of Eukaryotic Cells: Evidence and Research Implications for a Theory of the Origin and Evolution of Microbial, Plant, and Animal Cells on the Precambrian Earth.* New Haven, Ct.: Yale University Press, 1970.

Margulis, Lynn, and Karlene V. Schwartz. *Five Kingdoms: An Illustrated Guide to the Phyla of Life on Earth*. New York: W.H. Freeman and Co., 2000.

Margulis, Lynn, and Dorion Sagan. *Acquiring Genomes: A Theory of the Origins of Species*. New York: Basic Books, 2002.

Margulis, Lynn, et al., eds. *Chimeras and Consciousness: Evolution of the Sensory Self*. New York: W.H. Freeman and Co., 2006.

Maynard Smith, John and Eörs Szathmáry. *The Major Transitions in Evolution*. Oxford and New York: Oxford University Press, 1995.

———. *The Origins of Life: From The Birth of Life to the Origin of Language*. Oxford and New York: Oxford University Press, 1999.

Mayr, Ernst, and William B. Provine, eds. *The Evolutionary Synthesis: Perspectives on the Unification of Biology* (reprint). Cambridge, Mass.: Harvard University Press, 1998.

Moore, Raymond C., ed. *Treatise on Invertebrate Paleontology*. Lawrence, Kans.: Geological Society of America and Kansas University Press, 1968.

Morris, Simon Conway. *The Crucible of Creation: The Burgess Shale and the Rise of Animals*. Oxford and New York: Oxford University Press, 1998.

———. *Life's Solution: Inevitable Humans in a Lonely Universe*. Paperback, Cambridge: Cambridge University Press, 2005.

Nijhout, H. Frederik. *Insect Hormones*. Princeton, N.J.: Princeton University Press, 1994.

Novák, Vladimír J.A. *Insect Hormones*. Trans. Frome and London: Butler & Tanner Ltd., 1966.

Ohno, Susumu. *Evolution by Gene Duplication*. Berlin and New York: Springer-Verlag, 1970.

Popper, Karl Raimund. *The Logic of Scientific Discovery* (paperback reprint). London: Routledge, 2000.

Raff, Rudolph A. *The Shape of Life: Genes, Development, and the Evolution of Animal Form*. Chicago: University of Chicago Press, 1996.

Raff, Rudolph A., and Elizabeth C. Raff, eds. *Development as an Evolutionary Process: Proceedings of a Meeting Held at the Marine Biological Laboratory in Woods Hole, Massachusetts, August 23 and 24, 1985*. New York: A.R. Liss, 1987.

Ryan, Frank P. *Tuberculosis: The Greatest Story Never Told*. Bromsgrove: Swift Publishers, 1992.

———. *Virus X: Tracking The New Killer Plagues – Out of the Present and Into the Future*. London: HarperCollins, 1996.

———. *Darwin's Blind Spot: Evolution Beyond Natural Selection*. London: Thomson Learning, 2002.

———. *Virolution*. London: HarperCollins, 2009.

Sapp, Jan. *Evolution by Association: A History of Symbiosis*. Oxford and New York: Oxford University Press, 1994.

Sturtevant, Alfred Henry. *A History of Genetics*. New York: Harper & Row, 1965.

Syvanen, Michael and Clarence I. Kado, eds. *Horizontal Gene Transfer*. London: Chapman & Hall, 1998.

Valentine, James W. *On the Origin of Phyla*. Chicago: University of Chicago Press, 2004.

Villarreal, Luis P. *Viruses and the Evolution of Life*. Washington D.C.: American Society for Microbiology Press, 2005.

Wigglesworth, Vincent B. *Insects and the Life of Man: Collected Essays on Pure Science and Applied Biology*. London: Chapman & Hall, 1976.

——. *The Life of Insects*. London: Weidenfield and Nicholson, 1972.

——. *The Physiology of Insect Metamorphosis*. Cambridge: Cambridge University Press, 1954.

Williamson, Donald I. *Larvae and Evolution: Towards a New Zoology*. New York: Chapman & Hall, 1992.

——. *The Origins of Larvae*. Dordrecht and Boston: Kluwer Academic Publishers, 2003.

Willmer, Pat. *Invertebrate Relationships: Patterns in Animal Evolution*. New York and Cambridge: Cambridge University Press, 1990.

# Notes

Prologue: The Beautiful Mystery
1. Conrad, Nicholas J. "Palaeolithic Ivory Sculptures from Southwestern Germany and the Origins of Figurative Art." *Nature* 426 (18 December 2003): 830–2. See also Sinclair, Anthony. "Archaeology: Art of the ancients." *Nature* 426 (18 December 2003): 774–5. Dalton, Rex. "Lion man takes pride of place as oldest statue." *Nature* 425 (4 September 2003): 7.
2. Hammond, Norman. "Stone Age Artists Are Getting Older." *Times Online* (13 February 2006). Roach, John. "Ancient Figurines Found – from First Modern Humans?" *National Geographic News* (17 December 2003).
3. Wigglesworth, Vincent B. *Insects and the Life of Man* (London: Chapman and Hall, 1976), p. 168.

Part I
1. This is the frontispiece quote from *The Physiology of Insect Metamorphosis* by Vincent B. Wigglesworth (Cambridge: Cambridge University Press, 1954).

1. The Birth of an Idea
1. All direct quotes from Williamson are from my interviews and various communications.
2. Fernald, Russell D. "The Evolution of Eyes." *Karger Gazette* 64 (January 2001), http://www.karger.com/gazette/64/fernald/art_1_0.htm.

2. A Puzzle Wrapped in an Enigma
1. Tattersall, Walter Medley, and E.M. Sheppard. "Observations on the Bipinnaria of the Asteroid Genus Luidia." *Lancashire Sea-Fisheries Laboratory: James Johnstone Memorial Volume* (Liverpool: Liverpool University Press, 1934), pp. 35–61.
2. Haeckel, Ernst Heinrich. "Die Gastraea-Theorie, Die Phylogenetische Classification des Thierreichs und die Homologie der Keimblätter." *Jenaische Zeitschrift fur Naturwiss* 8 (1874): 1–55.

3. Balfour, Francis M. *A Treatise on Comparative Embryology*, vol. 2 (London: Macmillan & Co., 1880–1), p. 381.
4. Darwin, Francis, ed. *The Life and Letters of Charles Darwin, Including an Autobiographical Chapter* (New York: D. Appleton & Company, 1901), p. 426.
5. Ibid.
6. Hall, Brian K. "Balfour, Garstang and de Beer: The First Century of Evolutionary Embryology." *American Zoologist* 40 (5): 718–28. See also Hall, Brian K. and Marvalee Wake, eds. *The Origin and Evolution of Larval Forms* (San Diego: Academic Press, 1999).

3. First Experiments
1. Darwin, Charles. *The Descent of Man* (New York: Prometheus Books, 1998).
2. Barrington, E.J.W. "Metamorphosis in Lower Chordates." In Etkin, William and Lawrence I. Gilbert, eds. *Metamorphosis: a Problem in Developmental Biology* (New York: Appleton-Century-Crofts, 1968).
3. Meinertzhagen, Ian A. and Yasushi Okamura, "The Larval Ascidian Nervous System: The Chordate Brain from Its Small Beginnings." *Trends in Neurosciences* 24 (July 2001): 401–10.

4. The Price of Iconoclasm
1. Fell, Howard Barraclough. "The Direct Development of a New Zealand Ophiuroid." *Quarterly Journal of Microscopical Science* 82 (1941): 377–441.
2. In the following paper, Kirk uses the old term "sand-star" for what is now known as a brittle star.
3. Kirk, H.B. "Much Abbreviated Development of a Sand-star (*Ophionereis schayeri?*)" *Transactions of the New Zealand Institute* 48 (1916): 12–18.
4. Hyman, Libbie Henrietta. *The Invertebrates* vols. 1–5. (New York: McGraw-Hill, 1940–9).
5. Russo, A. "Embriologia dell' Amphiura Squamata." *Transactions of the Royal Academy of Naples* 2,5 (1891)
6. Williamson, Donald I. *The Origins of Larvae* (London and Boston: Kluwer Academic Publishers, 2003), p. 117.
7. Fell, Howard Barraclough. "Echinoderm Ontogeny." In Moore, Raymond C., ed. *Treatise on Invertebrate Paleontology* (Lawrence, Kans.: Geological Society of America and Kansas University Press, 1968), pp. 60–85.
8. Rowe, Francis W.E., Alan N. Baker, and Helen E.S. Clark. "The Morphology, Development and Taxonomic Status of Xyloplax Baker, Rowe & Clark (1986) (*Echinodermata concentricylcloidea*), with Description of a New Species." *Proceedings of the Royal Society of London B* 23 (May 1988): 431–59.
9. Schatt, Philippe, and Jean-Pierre Féral. "Completely Direct Development of *Abatus cordatus*, a Brooding Schizasterid (Echinodermata, Echinoidea) from

Kerguelen, with Description of Perigastrulation, a Hypothetical New Mode of Gastrulation." *Biological Bulletin* 190 (1996): 24–44.

10. Williamson, Donald I. *Larvae and Evolution: Towards a New Zoology* (New York: Chapman & Hall, 1992), preface.

5. Challenging the Tree of Life

1. Williamson, Donald I. "Incongruous Larvae and the Origin of Some Invertebrate Life-histories." *Progress in Oceanography* 19 (1987): 87–116.
2. Darwin, Charles. *On the Origin of Species by Means of Natural Selection* (London: Penguin, 1982), 1st ed., ch. 13.
3. Ibid., 6th ed., ch. 14.
4. Ibid., 6th ed., ch. 14.
5. Willmer, Pat. *Invertebrate Relationships: Patterns in Animal Evolution* (New York and Cambridge: Cambridge University Press, 1990), p. 123. See also p. 313.

6. 'This Is Impossible!'

1. Margulis, Lynn. Origin of Eukaryotic Cells: Evidence and Research Implications for a Theory of the Origin and Evolution of Microbial, Plant, and Animal Cells on the Precambrian Earth (New Haven, Ct.: Yale University Press, 1970).
2. Williamson, Donald I. *Larvae and Evolution: Towards a New Zoology* (New York: Chapman & Hall, 1992), foreword.
3. Darwin, Charles. *On the Origin of Species by Means of Natural Selection* (London: Penguin, 1982), 6th ed., ch. 14.
4. Interviews with Williamson. See also: Williamson, Donald I. *Larvae and Evolution.*
5. Williamson, Donald I. *Larvae and Evolution*, editor's comments.

8. The Evening of the Great Peacock

1. Fabre, Jean-Henri. 'The Life of the Caterpillar.' In *Souvenirs Entomologiques.* English translation by Alexander Teixeira de Mattos, London: Hodder and Stoughton, 1912.
2. Darwin, C. *The Descent of Man* (New York: Prometheus Books, 1998).
3. Alcock, John. *In a Desert Garden: Love and Death Among the Insects* (Tucson: University of Arizona Press, 1999), p. 38.
4. Letter from Charles Darwin to Jean-Henri Fabre of January 31 1880. http://www.e-fabre.com/en/virtual_library/correspondence/darwinletter.htm.
5. Legros, Georges V. *Fabre, Poet of Science,* trans. Bernard Miall (Oxford, Miss.: Project Gutenberg Literary Archive Foundation, 2002).

9. The Science of Life
 1. Wigglesworth, Vincent B. "Metamorphosis, Polymorphism, Differentiation." *Scientific American* 200 (1959): 100–10. See also: Wigglesworth, Vincent B. "Some Memories: Interview with Sir Vincent Wigglesworth, Fellow and Emeritus Quick Professor of Biology." *The Caian* (1979): 30–44.
 2. Wigglesworth, Vincent B. "Metamorphosis, Polymorphism, Differentiation."
 3. Wigglesworth, Vincent B. *Insects and the Life of Man: Collected Essays on Pure Science and Applied Biology* (London: Chapman & Hall, 1976), ch. 12, p. 138, *et seq*.
 4. Ibid.
 5. At this time Fabre had been translated into English, and his books were popular among the natural history-minded English middle classes. In a subsequent interview with Wigglesworth's son, Professor Jonathan Wigglesworth, he confirmed that his father had read and was abundantly familiar with Fabre. Indeed, Wigglesworth in turn gave a book by Fabre to his son as a child. From this interview I also gathered that Wigglesworth was a practicing Christian, but, like most British Christian scientists, he saw no contradiction in accepting Darwin's concept of evolution.
 6. Wigglesworth, Vincent B. *The Life of Insects* (London: Weidenfeld and Nicolson, 1964), p. 1.
 7. Wigglesworth, Vincent B. *Insects and the Life of Man*, p. 138.
 8. Ibid., p. 139.
 9. Mortimer, Philip P. "The Control of Yellow Fever: A Centennial Account." *Microbiology Today*, 29 (2002): 24–6.
 10. Wigglesworth, Vincent B. "Some Memories", p. 39.
 11. Edwards, John S. "Sir Vincent Wigglesworth and the Coming of Age of Insect Development." *International Journal of Developmental Biology* 42 (1998): 471–3.

10. Elementary Questions and Deductions
 1. Wigglesworth, Vincent B. "The Physiology of the Cuticle and of Ecdysis in *Rhodnius prolixus* (Triatomidae. Hemiptera); with Special Reference to the Function of the Oenocytes and the Dermal Glands." *Quarterly Journal of Microscopical Science* 76 (1933): 269–318.
 2. Ibid.
 3. Wigglesworth, Vincent B. *The Life of Insects* (London: Weidenfeld and Nicolson, 1964), p. 10.
 4. Engel, Michael S. and David A. Grimaldi. "A New Light Shed on the Oldest Insect." *Nature* 427 (2004): 627–30. See also: Muir, Hazel. "Earliest Ever Insect Fossil Springs Winged Surprise." *New Scientist* (14 February 2004): 9.

11. The Phoenix in Its Crucible
 1. Wigglesworth, Vincent B. *The Physiology of Insect Metamorphosis* (Cambridge: Cambridge University Press, 1954), p. 8.
 2. Ibid.
 3. Ibid., p. 5.
 4. See for example: Bennet-Clark, Henry C. "The Effect of Air Resistance on the Jumping Performance of Insects." *Journal of Experimental Biology* 82 (1979): 105–21. Bennet-Clark, Henry C. "A Model of the Mechanism of Sound Production in Cicadas." *Journal of Experimental Biology* 173 (1992): 123–53. Bennet-Clark, Henry C. "The Stability of Swivel Wing Supersonic Aircraft." *Nature* 239 (1972): 451–2.
 5. There is a detailed study of Lehmann in my book, *The Forgotten Plague*.
 6. Interviews and communications with Henry Bennet-Clark.
 7. Interview with Simon Maddrell.

12. Two Souls in One Body
 1. Wigglesworth, Vincent B. "The Physiology of the Cuticle and of Ecdysis in *Rhodnius prolixus* (Triatomidae. Hemiptera); with Special Reference to the Function of the Oenocytes and the Dermal Glands." *Quarterly Journal of Microscopical Science* 76 (1933).
 2. I personally favour a much commoner and more prosaic explanation of irritable bowel, which is often stress related, and perhaps complicated by diverticular disease – with perhaps his terminal illness brought about by one or more of the unpleasant long-term complications, such as stricture development or diverticular abscess and peritonitis. Another altogether commonplace and plausible explanation for his death, given his age and symptoms, might be a slow-growing colonic cancer.
 3. Wigglesworth, Vincent B. "Factors Controlling Moulting and 'Metamorphosis' in an Insect." *Nature* 131 (1934): 725–6.
 4. Kopeč, Stefan. "Studies on the Necessity of the Brain for the Inception of Insect Metamorphosis." *Biological Bulletin* 42 (1922): 322–42.
 5. Kopeč, Stefan. "Über die Entwicklung der Insecten unter dem Einfluss der Vitaminzugabe." *Biologica Generalis* 3 (1927): 375–84.

13. Bizarre Extrapolations
 1. Gilbert, Scott F. "Do Humans Undergo Metamorphosis?" *Developmental Biology*, 8th ed., http://8e.devbio.com/article.php?id=9.
 2. See for example: Fevold, Harry L., Frederick L Hisaw, and S.L. Leonard, "The Gonad Stimulating and the Luteinizing Hormones of the Anterior Lobe of the Hypophesis." *American Journal of Physiology* 97 (1931): 291–301.
 3. Wigglesworth's subsequent recognition of the importance of Kopeč's work led to a belated recognition of the Polish scientist, who enjoyed a brief fame

as a pioneer of endocrinology before his untimely death in war-torn Poland in 1941.

4. Wigglesworth, Vincent B. "The Physiology of Ecdysis in *Rhodnius prolixus* (Hemiptera): Factors Controlling Moulting and 'Metamorphosis' in an Insect." *Quarterly Journal of Microscopical Science* 77 (1934): 191–222.

5. Ibid., 197. See also Wigglesworth, Vincent B. "Metamorphosis, Polymorphism, Differentiation." *Scientific American* 200 (1959).

6. Wigglesworth, Vincent B. "The Physiology of Ecdysis", p. 211.

7. Wigglesworth, Vincent B. "The Function of the Corpus Allatum in the Growth and Reproduction of *Rhodnius prolixus* (Hemiptera)." *Quarterly Journal of Microscopical Science* 79 (1936): 91–121.

8. Wigglesworth, Vincent B. "The Determination of Characters at Metamorphosis in *Rhodnius prolixus* (Hemiptera)." *Journal of Experimental Biology* 17 (1940): 201–22.

9. Ibid., 221.

10. Bounhiol, Jean-Jacques. "Recherches expérimentales sur le déterminisme de la metamorphose chez le les Lépidoptères." *Bulletin Biologique de la France et de la Belgique,* suppl. 24 (1938): 1–199. See also the biography of Bounhiol at Lamy, M., and M. Delsol. "Jean-Jacques Bounhiol, 1905–1979." *Annales d'Endocrinologie* 41 (1980): 153–6. Plagge, Ernst. "Weitere Untersuchungen über das Verpuppungshormon bei Schmetterlingen." *Biol. Zentralbl.* 58 (1938): 1–12.

11. Wigglesworth, Vincent B. *The Physiology of Insect Metamorphosis* (Cambridge: Cambridge University Press, 1954), p. 26.

12. Hachlow, V. "Zur Entwicklungsmechanik der Schmetterlinge." *Roux Arch. EntwMech. Organ.* 125 (1931): 26–49.

13. Fukuda, Soichi. "Hormonal Control of Moulting and Pupation in the Silkworm." *Proceedings of the Imperial Academy of Japan* 16 (1940): 417–20. Fukuda, Soichi. "Role of the Prothroracic Gland in Differentiation of the Imaginal Characters in the Silkworm Pupa." *Annotationes Zoologicae Japonenses* 20 (1941): 9–13. Fukuda, Soichi. "The Hormonal Mechanism of Larval Moulting and Metamorphosis in the Silkworm." *Journal of the Faculty of Science, Tokyo University* 6 (1944): 477–532.

14. Toyama, Kametaro. "Contributions to the Study of Silk-worms: On the Embryology of the Silk-worm." *Journal of the College of Agriculture, Imperial University of Tokyo* 5 (1902): 73–118. Ke, O. "Morphological Variation of the Prothoracic Gland in the Domestic and Wild Silkworm." *Bulletin of the Faculty of Sciences Terkult., Kyushu Imperial University* 4 (1930): 12–21.

15. Interview with Chris Curtis. See also: Clark, Catherine, M. and James Macalister Mackintosh. *The School and the Site: A Historical Memoir to Celebrate the Twenty-fifth Anniversary of the School* (London: H.K Lewis & Co. Ltd., 1954), which contains images of the bomb damage.

16.  Interview with Jonathan Wigglesworth.
17.  Interview with Henry Bennet-Clark.

14.  Assembling the Jigsaw Puzzle
 1.  Williams, Carroll M., and Austin H. Clarke. "Records of *Argynnis diana* and
     of Some Other Butterflies from Virginia." *Journal of the Washington Academy of
     Sciences* 27 (1937): 209–13.
 2.  Pappenheimer, Alwin Max. "Carroll Milton Williams: December 2, 1916
     – October 11, 1991." *Biographical Memoirs* 68 (1995): 413–33.
 3.  Ibid.
 4.  Gordon, N. "Juvenile Hormone Research Reaches into Fantastic Realms."
     *Boston Sunday Herald* (2 August 1959). Article kindly provided by the Harvard
     Archives.
 5.  Cousin, Germaine. "Étude expérimentale de la diapause des insects." *Bul-
     letin Biologique de la France et de la Belgique* suppl. 15 (1932): 1–341.
 6.  Williams, Carroll M. "Continuous Anesthesia for Insects." *Science* 103 (1946):
     57.
 7.  Williams, Carroll M. "Physiology of Insect Diapause: The Role of the Brain
     in the Production and Termination of Pupal Dormancy in the Giant Silk-
     worm, *Platysamia cecropia*." *Biological Bulletin* 90 (1946): 234–43.
 8.  Ibid.
 9.  Ibid.
10.  Williams, Carroll M. "Physiology of Insect Diapause, II: Interaction between
     the Pupal Brain and Prothoracic Glands in the Metamorphosis of the Giant
     Silkworm, *Platysamia Cecropia*." *Biological Bulletin* 93 (1947): 89–98.
11.  Ibid.
12.  Ibid. One gland was not enough – it was too difficult to dissect it out in its
     entirety.
13.  Ibid.
14.  Williams, Carroll M. "The Physiology of Insect Diapause, III: The Pro-
     thoracic Glands in the Cecropia Silkworm, with Special Reference to Their
     Significance in Embryonic and Postembryonic Development." *Biological
     Bulletin* 94 (1948): 60–5. Williams, Carroll M. "The Physiology of Insect
     Diapause, IV: The Brain and Prothoracic Glands as an Endocrine System
     in the Cecropia Silkworm." *Biological Bulletin* 103 (1952): 120–38. See also,
     Williams, Carroll M. "The prothoracic glands of insects in retrospect and in
     prospect." *Biological Bulletin* 97 (1949): 111–4.
15.  Pappenheimer, Alwin Max. "Carroll Milton Williams."
16.  Interview with Lynn Riddiford and James Truman.
17.  Interview with Simon Maddrell.

15. Ecology's Magic Bullet

1. Butenandt, Adolf, and Peter Karlson. "Uber die isolierung eines Metamorphose-hormones der Insekten in Kristallisierter Form." *Z Naturforsch* 9b (1954): 389–91. See also: Karlson, Peter, et al. "Zur chemie des ecdysons." *Justus Liebigs Annalen der Chemie* 662 (1963): 1–20.

2. Pappenheimer, Alwin Max. "Carroll Milton Williams: December 2, 1916 – October 11, 1991." *Biographical Memoirs* 68 (1995): 413–33.

3. Williams, Carroll M. "The Juvenile Hormone of Insects." *Nature* 178 (1956): 212–3.

4. McElheny, Victor K. "Key Found to Insect Self-curbs." *Boston Globe* (27 December 1967). Kindly provided by the Harvard Archives.

5. Pappenheimer, Alwin Max. "Carroll Milton Williams. See also Riddiford, Lynn M., and Carroll M. Williams. "Volatile Principle from Oak Leaves: Role in Sex Life of the Polyphemus Moth." *Science* 155 (1967): 589–90.

6. Williams, Carroll M. "The Juvenile Hormone, I: Endocrine Activity of the Corpora Allata of the Adult Cecropia Silkworm." *Biological Bulletin* 116 (1961): 323–38. Williams, Carroll M. "The Juvenile Hormone, II: Its Role in the Endocrine Control of Molting, Pupation and Adult Development in the Cecropia Silkworm." *Biological Bulletin* 121 (1961): 572–85. Williams, Carroll M. "The Juvenile Hormone, III: Its Accumulation and Storage in the Abdomen of Certain Male Moths." *Biological Bulletin* 124 (1963): 355–67. Williams, Carroll M., and John H. Law. "The Juvenile Hormone, IV: Its Extraction, Assay, and Purification." *Journal of Insect Physiology* 11 (1965): 569–80. Williams, Carroll M., and Karel Slama. "Juvenile Hormone Activity for the Bug *Pyrrhocoris apterus*." *Proceedings of the National Academy of Sciences* 54 (1965): 411–14.

7. In fact the hormone is inactive in the form it first enters the blood and has to be converted into its active form, 20-hydroxyecdysterone, in the peripheral tissues. This introduces an extra layer of complexity to its action – prior cellular activation.

8. Williams, Carroll M., and Andrew Spielman. "Lethal Effects of Synthetic Juvenile Hormone on Larvae of the Yellow Fever Mosquito, *Aedes Egypti*." *Science* 154 (1966): 1043–44.

9. Williams, Carroll M. "Third-generation Pesticides." *Scientific American* 217 (1967): 13–17.

10. Williams, Carroll M., and Lynn M. Riddiford. "The Effects of Juvenile Hormone Analogues on the Embryonic Development of Silkworms." *Proceedings of the National Academy of Sciences* 57 (1967): 595–601.

11. McElheny, Victor K. "Key Found to Insect Self-curbs."

12. Yapabandara, Amara M, G, M, Christopher F. Curtis, et al. "Control of

Malaria Vectors with Insect Growth Regulator Pyriproxifen in a Gem-mining Area of Sri Lanka." *Acta Tropica* 80 (2001): 265–76.

13. Nijhout, H. Frederik. "Hormonal Control in Larval Development and Evo-lution – Insects." In Hall, Brian K., and Marvalee H. Wake, eds. *The Origin and Evolution of Larval Forms* (San Diego: Academic Press, 1999).


16. On the Steps of York Minster

1. Williamson, Donald I. *Larvae and Evolution: Towards a New Zoology* (New York: Chapman & Hall, 1992).


17. The First Genetic Testing

1. Personal communication with Mike Hart. Richard Strathmann's research interest viewpoint is derived from his URL at the University of Washington.

2. Strathmann, Richard R. "Larvae and Evolution: Towards a New Zoology." *Quarterly Review of Biology* 68 (1993): 280–2.

3. Personal communication with Michael Hart.

4. Williamson, Donald I. *Larvae and Evolution: Towards a New Zoology* (New York: Chapman & Hall, 1992), p.183.

5. Personal communication with Michael Hart.

6. Hart, Michael W. "Testing Cold Fusion of Phyla: Maternity in a Tunicate x Sea Urchin Hybrid Determined from DNA Comparisons." *Evolution* 50 (1996): 1713–8.


18. Novelties Can Arise

1. Raff, Rudolf A. "Constraint, Flexibility, and Phylogenetic History in the Evolution of Direct Development in Sea Urchins." *Developmental Biology* 119 (1987): 6–19.

2. Raff, Elizabeth C., Ellen M. Popodi, et al. "A Novel Ontogenic Pathway in Hybrid Embryos between Species with Different Modes of Development." *Development* 126 (1999): 1937–45.

3. Raff, Rudolf A., Louis Herlands, et al. "Evolutionary Modification of Echi-noid Sperm Correlates with Developmental Mode." *Development, Growth and Differentiation* 32 (1990): 283–91.

4. Rahman, M. Aminur, Tsuyoshi Uehara, and John S. Pearse. "Hybrids of Two Closely Related Tropical Sea Urchins (Genus Echinometra): Evidence against Postzygotic Isolating Mechanisms." *Biological Bulletin* 200 (2001): 97–106.

5. Uehara, Tsuyoshi, H. Asakura, and Yuji Arakaki. "Fertilization block-age and hybridisation among species of sea urchins." In Hoshi, Motonori and Okitsugu Yamashita, eds. *Advances in Invertebrate Reproduction, 5: Fifth*

*International Congress of Invertebrate Reproduction, Nagoya, Japan, July 23–28, 1989* (Amsterdam and New York: Elsevier, 1990).

6. Raff, Elizabeth C., and Ellen M. Popodi, et al. "A Novel Ontogenic Pathway."

7. Willmer, Pat. *Invertebrate Relationships: Patterns in Animal Evolution* (New York and Cambridge: Cambridge University Press, 1990), p. 117, *et seq.*

8. Personal communication with Mark Nielsen.

9. Nielsen, Mark G., Keen A. Wilson, et al. "Novel Gene Expression Patterns in Hybrid Embryos between Species with Different Modes of Development." *Evolution and Development* 2 (2000): 133–44.

10. Komatsu, Meiko, and Takako Chimura. "Development of Hybrid Embryos between Species in Seastars and Sea Urchins (abstract)." *Zoological Science*, suppl. 18 (2001): 80.

11. Komatsu, Meiko, Takako Chimura, and Yuki Yamazaki. "Larval Development and Genetic Detection of Echinoderm Hybrids using RAPD-PCR Technique (abstract)." *Zoological Science* 19 (2002): 1458.

12. Personal communication with Kaori Wakabashi.

13. I devoted two chapters to the growing field of hybridisation in my book *Virolution.*


19. A New Life Form

1. Interview with Sebastian Holmes (2002).

2. Personal communication with Lynn Margulis.

3. Interview with Holmes (2002).

4. Holmes, Sebastian P., Donald, I. Williamson, and Nic Boerboom. "Phylogenetic Hybridisation: A Source of Genetic Novelty in Contravention of Dollo's Law?" Presented as a poster at the Thirty-eighth European Marine Biology Symposium in Alverio, Portugal (June 2003). See also: Williamson, Donald I. "Larval Transfer: Evolution by Hybridisation." In Margulis, Lynn, Celeste A. Asikainen, et al., eds. *Chimeras and Consciousness: Evolution of The Sensory Self* (New York: W.H. Freeman and Co., 2006).

5. Ryan, Frank. "An Alternative Approach to Medical Genetics Based on Modern Evolutionary Biology, 5: Epigenetics and Genomics." *Journal of the Royal Society of Medicine* 102 (2009): 530–7.

6. Gilbert, Scott F. *Developmental Biology*, 7th ed. (Sunderland, Mass.: Sinauer Associates, 2003), p. 33.

7. Bengston, Stefan, and Yue Zhao. "Fossilized Metazoan Embryos from the Earliest Cambrian." *Science* 277 (1997): 1645–8.

8. Rasmussen, Birger, Stefan Bengston, et al. "Discoidal Impressions and Trace-like Fossils More than 1200 Million Years Old." *Science* 296 (2002): 1112–5.

9. Chen, Jun-Yuan, Paola Oliveri, et al. "Precambrian Animal Diversity: Putative Phosphatized Embryos from the Doushantuo Formation of China." *Proceedings of the National Academy of Sciences* 97 (2000): 4457–62.

10. Xiao, Shuhai, Xunlai Yuan, et al. "Microscopic Carbonaceous Compression in a Terminal Proterozoic Shale: A Systematic Reassessment of the Miaohe Biota, South China." *Journal Paleontology* 76 (2002): 347–76.

20. The Puzzle of the Hornworm Brain

1. Truman, James W., and Lynn M. Riddiford. "The Origin of Insect Metamorphosis." *Nature* 401 (1999): 447–52.

2. Riddiford, Lynn M., and Carroll M. Williams. "Chemical Signalling between Polyphemus Moths and between Moths and Host Plant." *Science* 156 (1967): 541.

3. Goodman, Corey S. and Bridget C. Coughlin. "The Evolution of Evo-devo Biology." *Proceedings of the National Academy of Sciences* 97 (2000): 4424–5.

4. Carroll, Sean B. "Evolution at Two Levels: On Genes and Form." *PLoS Biology* 3 (2005): 245.

5. Nielsen, Claus. "The Origin of Metamorphosis." *Evolution & Development* 2 (2000): 127–9.

6. Jägersten, Gösta. *Evolution of the Metazoan Life Cycle: A Comprehensive Theory* (London and New York: Academic Press, 1972).

7. Kristensen, Niels P. "The Phylogeny of Hexapod 'Orders': A Critical View of Recent Accounts." *Zoology Systematic Evolutionsforschung* 13 (1975): 1–44.

8. Judson, Horace F. *The Eighth Day of Creation* (London: Penguin, 1995), p. 194.

9. Interview with James Truman and Lynn Riddiford.

10. Horodyski, Frank M., Lynn M. Riddiford, and James W. Truman. "Isolation and Expression of the Eclosion Hormone Gene from the Tobacco Hornworm, *Manduca sexta.*" *Proceedings of the National Academy of Sciences* 86 (1989): 8123–7.

11. Rankin, Mary-Ann, and Lynn M. Riddiford. "The Hormonal Control of Migratory Flight in *Oncopeltus fasciatus*: The Effects of Corpus Cardiacum, Corpus Allatum and Starvation on Migration and Reproduction." *General and Comparative Endocrinology* 33 (1977): 309–21.

12. Personal communication with James Truman.

13. Prokop, Andreas, and Gerhard M. Technau. "The Origins of Postembryonic Neuroblasts in the Ventral Nerve Cord of *Drosophila melanogaster.*" *Development* 111 (1991): 79–88.

21. Aristotle or Darwin?

1. Berlese, Antonio. "Intorno alle metamorfosi degli insetti." *Redia* 9 (1913): 121–36.

2. Novák, Vladimír J.A. *Insect Hormones* (Rome and London: Butler & Tanner Ltd., 1966), p. 137.

3. Ibid., p. 142, *et seq*. Hinton, Howard E. "On the Origin and Function of the Pupal Stage." *Transactions of the Royal Entomological Society, London* 99 (1948): 395–409. Interview with Henry Bennet-Clark.

4. It is interesting to note that Darwin, and thus Wigglesworth, followed the general interpretation first proposed by the seventeenth-century Dutch naturalist, Jan Swammerdam. See, Novák, Vladimír J.A. *Insect Hormones*, p. 149, *et seq*.

5. Truman, James W., and Lynn M. Riddiford. "Endocrine Insights into the Evolution of Metamorphosis in Insects." *Annual Review of Entomology* 47 (2001): 467–500.

6. Ibid., p. 470. See also, Corbet, P.S. "The Immature Stages of the Emperor Dragonfly, *Anax imperator* Leach (Odonata: Aeshnidae)." *Entomologist's Gazette* 6 (1955): 189–97.

7. Erezyilmaz, Deniz, Lynn M Riddiford, and James W. Truman. "The Pupal-specific *Broad* Directs Progressive Morphogenesis in a Direct-developing Insect." *PNAS* (*Proceedings of the National Academy of Sciences*), in press.

22. Cues and Common Links

1. Judson, Horace F. The Eighth Day of Creation: Makers of the Revolution in Biology (London: Penguin, 1995), p. 218.

2. Cameron, Andrew, Gregory Mahairas, et al. "A Sea Urchin Genome Project: Sequence Scan, Virtual Map, and Additional Resources." *Proceedings of the National Academy of Sciences* 97 (2000): 9514–8.

3. Davidson, Eric H. *Genomic Regulatory Systems: Development and Evolution* (San Diego: Academic Press, 2001), p. 19.

4. Carroll, Sean B. "Evolution at Two Levels: On Genes and Form." *PLoS Biology* 3 (2005): 1159–66. I should express my thanks to Alan Chalk who pointed out this article to me.

5. Gilbert, Scott F. "Environmental Regulation of Normal Development." *Developmental Biology* (Sunderland Mass.: Sinauer Associates, 2000).

6. Nijhout, H. Frederik. *Insect Hormones* (Princeton, N.J.: Princeton University Press, 1994).

7. Jackson, Daniel, et al. "Ecological Regulation of Development: Induction of Marine Invertebrate Metamorphosis." *International Journal of Developmental Biology* 46 (2002): 679–86.

8. For a general discussion of environmental regulation of animal development, see chapter 22 in Gilbert, Scott F. *Developmental Biology* 7th ed. (Sunderland, Mass.: Sinauer Associates, 2003). For a specific discussion of the bacterial symbionts in animal development see, McFall-Ngai, Margaret J. "Unseen

Forces: The Influence of Bacteria on Animal Development." *Developmental Biology* 242 (2002): 1–14. McFall-Ngai, Margaret J., and Edward G. Ruby. "Symbiont Recognition and Subsequent Morphogenesis as Early Events in an Animal-bacterial Mutualism." *Science* 254 (1991): 1491–4. Montgomery, Mary K., and Margaret J. McFall-Ngai, "The Inductive Role of Bacterial Symbionts in the Morphogenesis of a Squid Light Organ." *American Zoologist* 35 (1995): 372–80.

9. Grant-Downton, Robert T., and Hugh G. Dickinson. "Epigenetics and Its Implications for Plant Biology 2 – The 'Epigenetic Epiphany': Epigenetics, Evolution and Beyond." *Annals of Botany* 97 (2006): 11–27.

10. Beerman, Wolfgang. "Chromosomeren Konstanz und Spezifische Modifikation der Chromosomerenstruktur in der Entwicklung und Organ differentzierung von *Chironomus tentans*." *Chromosoma* 5 (1952): 139–384.

11. Clever, Ulrich, and Peter Karlson. "Induktion von Puff-Veranderungen in den Speicheldrüsenchromosomen von *Chironomus tentans* durch ecdysone." *Experimental Cell Research* 20 (1960): 623–6.

12. Lezzi, Markus. "Chromosome Puffing: Supramolecular Aspects of Ecdysone Action." Gilbert, Lawrence I., Tata Jamshed R., *et al* eds. *Metamorphosis: Postembryonic Reprogramming of Gene Expression in Amphibian and Insect Cells.* San Diego: Academic Press, 1996.

13. Cherbas, Peter, and Lucy Cherbas. "Molecular Aspects of Ecdysteroid Hormone Action." Gilbert, Lawrence I., Jamshed R. Tata, et al., eds. *Metamorphosis: Postembryonic Reprogramming of Gene Expression in Amphibian and Insect Cells* (San Diego: Academic Press, 1996).

14. Gilbert, Lawrence I., Jamshed R. Tata, et al., eds. *Metamorphosis*, p. xv.

23. A Tale in a Tail

1. Barrington, E.J.W. "Metamorphosis in Lower Chordates." In Etkin, William, and Lawrence I. Gilbert, eds. *Metamorphosis: a Problem in Developmental Biology* (New York: Appleton-Century-Crofts, 1968).

2. We might also recall that there are adult forms, known as larvaceans, within the same phylum of the urochordates that are tailed tadpoles, possess gonads and which independently feed and reproduce.

3. Barrington, E.J.W. "Metamorphosis in Lower Chordates."

4. Ibid.

5. Ibid.

6. Jackson, Daniel, et al. "Ecological Regulation of Development: Induction of Marine Invertebrate Metamorphosis." *International Journal of Developmental Biology* 46 (2002).

7. Woods, Rick G., et al. "Gene Expression during Early Ascidian Metamorphosis Requires Signalling by Hemps, an EGF-like Protein." *Development* 131 (2004): 2921–33.

8. Patricolo, Eleanora, Matteo Cammarata, and Paolo D'Agati. "Presence of Thyroid Hormones in Ascidian Larvae and Their Involvement in Metamorphosis." *Journal of Experimental Zoology* 290 (2001): 426–30.

9. Patricolo, Eleanora, et al. "Organometallic Complexes with Biological Molecules, XVI: Endocrine Disruption Affects of Tributyltin(IV) Chloride on Metamorphosis of the Ascidian Larva." *Applied Organometallic Chemistry* 15 (2001): 916–23.

10. Carosa, Eleanora, Albertina Fanelli, et al. "*Ciona intestinalis* nuclear receptor 1: a member of steroid/thyroid hormone receptor family." *Proceedings of the National Academy of Sciences* 95 (1998): 11152–57. See also: Devine, Christine, Veronica F. Hinman, and Bernie M. Degnan. "Evolution and Developmental Expression of Nuclear Receptor Genes in the Ascidian *Herdmania*." *International Journal of Developmental Biology* 46 (2002): 687–92.

11. Johnson, Leland G. "Stage-dependent Thyroxine Effects on Sea Urchin Development." *New Zealand Journal of Marine and Freshwater Research* 32 (1998): 531–6. Heyland, Andreas, Adam Reitzel, and Jason Hodin. "Thyroid Hormones Determine Developmental Mode in Sand Dollars (Echinodermata: Echinoidea)." *Evolution and Development* 6 (2004): 382–92.

12. Youson, John H. "Is Lamprey Metamorphosis Regulated by Thyroid Hormone?" *American Zoologist* 37 (1997): 441–60.

13. Murr, E., and Sklower, A. "Untersuchungen über der inkretorischen Organe der Fische, I: Das Verhalten der schilddruse in der Metamorphose des Aales." *Zeitschrift für vergleichende Physiologie* 7 (1928): 279–88.

24. Of Frogs and Their Relatives

1. An entertaining as well as usefully historical account of these early tetrapods is to be found in Gould, Stephen Jay. *Wonderful Life: The Burgess Shale and the Nature of History* (London: Penguin, 1991).

2. Clack, Jennifer A. "Getting a Leg Up on Land." *Scientific American* (December 2005): 80–7. Holmes, Bob. "Meet Your Ancestor: The Fish that Crawled." *New Scientist* (9 September 2006): 35–9. McLeod, Myles. "One Small Step for Fish, One Giant Leap for Us." *New Scientist* (19 August 2000): 28–32.

3. Dent, James Norman. "Survey of Amphibian Metamorphosis." In Etkin, William, and Lawrence I. Gilbert, eds. *Metamorphosis: A Problem in Developmental Biology* (New York: Appleton-Century-Crofts, 1968).

4. Etkin, William, and Lawrence I. Gilbert, eds. *Metamorphosis.* Gilbert, Lawrence I., Jamshed R. Tata, et al., eds. *Metamorphosis: Postembryonic Reprogramming of Gene Expression in Amphibian and Insect Cells* (San Diego: Academic Press, 1996). See in particular ch. 19, Yoshizato, Katsutoshi, "Cell Death and Histolysis in Amphibian Tail during Metamorphosis."

5. Kaltenbach, Jane C. "Endocrinology of Amphibian Metamorphosis." Gilbert, Lawrence I., Jamshed R. Tata, et al., eds. *Metamorphosis*.

6. See, for example, Kanamori, Akira, and Donald D. Brown. "The Analysis of Complex Developmental Programmes: Amphibian Metamorphosis." *Genes to Cells* 1 (1996): 429–35.

7. See Carnegie Institution, Department of Embryology, Staff Member biography page, http://www.ciwemb.edu/labs/brown/index.php. See also: Cai, Liquan, and Donald D. Brown. "Expression of Type II Iodothyronine Deiodinase Marks the Time that a Tissue Responds to Thyroid Hormone-induced Metamorphosis in *Xenopus laevis*." *Developmental Biology* 266 (2004): 87–95. See also: Marsh-Armstrong, Nick, Liquan Cai, and Donald D. Brown. "Thyroid Hormone Controls the Development of Connections between the Spinal Cord and Limbs during *Xenopus laevis* Metamorphosis." *PNAS* (*Proceedings of the National Academy of Sciences*) 101 (2004): 165–70.

8. Gehring, Walter J. *Master Control Genes in Development and Evolution: The Homeobox Story* (New Haven, Ct.: Yale University Press, 1998).

9. Ferrier, David E. K., and Carolina Minguillón. "Evolution of the Hox/ParaHox Gene Clusers." *International Journal of Developmental Biology* 47 (2003): 605–11.

25. To Be Human

1. Bronowski, Jacob. *The Ascent of Man* (BBC and Time-Life Films, 1973).

2. Gilbert, Scott F. "Do Humans Undergo Metamorphosis?" *Developmental Biology*, 8th ed., http://8e.devbio.com/article.php?id=9.

3. Gilbert, Scott F., and David Epel. *Ecological Developmental Biology: Integrating Epigenetics, Medicine, and Evolution* (Sunderland, Mass.: Sinauer Associates, 2009), p. 221–9.

4. See, for instance, http://en.wikipedia.org/wiki/Puberty

5. Gilbert, Scott F. "Childhood as a New Portion of the Human Life Cycle." *Developmental Biology*, 8th ed., http://www.devbio.com/article.php?ch=2&id=8.

6. Mohun, Tim, and Jim Smith. "Of Frogs and Men." *Mill Hill Essays: The National Institute of Medical Research*, http://www.nimr.mrc.ac.uk/mill-hill-essays/of-frogs-and-men.

7. Ryan, Frank. *Virolution* (London: HarperCollins, 2009).

8. Ryan, Frank. "I Virus." *New Scientist* (30 January 2010): 32–5. See also: Ryan, Frank. "An Alternative Approach to Medical Genetics based on Modern Evolutionary Biology, 2: Retroviral Symbiosis." *Journal of the Royal Society of Medicine* 102 (2009): 324–31.

9. Ryan, Frank. "Genomic Creativity and Natural Selection: A Modern Synthesis." *Biological Journal of the Linnean Society* 88 (2006): 655–72.

10. Gilbert, Scott F. "Do Humans Undergo Metamorphosis?" See also Bogin,

Barry. "Evolutionary Hypotheses for Human Childhood." *Yearbook of Physical Anthropology* 40 (1997): 63–89.

11. Rose, Stephen. *Lifelines: Biology Beyond Determininsm* (Oxford and New York: Oxford University Press, 1998).


26. The Sentient Brain

1. Zoeller, R. Thomas. "Editorial: Local Control of the Timing of Thyroid Hormone Action in the Developing Human Brain." *Journal of Clinical Endocrinology and Metabolism* 89 (2004): 3114–6.

2. Kilby, Mark D., and Shiao Chan. "Thyroid Hormone and Central Nervous System Development." *Journal of Endocrinology* 165 (2000): 1–8. Vulsma, Thomas, Margereth H. Gons, and Jan J. M. de Vijlder. "Maternal-fed Transfer of Thyroxine in Congenital Hypothyroidism Due to a Total Organifiction Defect of Thyroid Agenesis." *New England Journal of Medicine* 321 (1989): 13–6. Contempré, Bernard, Eric Jauniaux, et al. "Detection of Thyroid Hormones in Human Embryonic Cavities during the First Trimester of Pregnancy." *Journal of Clinical Endocrinol & Metabolism* 77 (1993): 1710–22.

3. Bernal, Juan, and Fredrika Pekonen. "Ontogenesis of the Nuclear 3.5.3'-triiodothyroxine Receptor in the Human Fetal Brain." *Endocrinology* 114 (1984): 677–9.

4. Bradley, D.J., H.C. Towle, and W.S. Young. "Spatial and Temporal Expression of Alpha- and Beta-thyroid Hormone Receptor mRNAs, including the Neta-two Subtype, in the Developing Mammalian Nervous System." *Journal of Neuroscience* 12 (1992): 2288–302.

5. Dowling, Amy L. S., Gabriel U. Martz, et al. "Acute Changes in Maternal Thyroid Hormone Induce Rapid and Transient Changes in Gene Expression in Fetal Rat Brain." *Journal of Neuroscience* 20 (2000): 2255–65.

6. Becker, Kathryn B., Kristen C. Stephens, et al. "The Type 2 and Type 3 Iodothyronine Deiodinases Play Important Roles in Coordinating Development in *Rana catesbeiana* Tadpoles." *Endocrinology* 138 (1997): 2989–97.

7. Marsh-Armstrong, Nick, Hauchi Huang, et al. "Asymmetric Growth and Development of the *Xenopus laevis* Retina during Metamorphosis Is Controlled by Type III Deiodinase." *Neuron* 24 (1999): 871–78.

8. See chapter 26, note 1, as well as Clancy, Barbara, Richard B. Darlington, and Barbara L. Finlay. "Translating Developmental Time across Mammalian Species." *Neuroscience* 105 (2001): 7–17.

9. Kester, Monique H. A., Raquel Martinez de Mena, et al. "Iodothyronine Levels in the Human Developing Brain: Major Regulatory Roles of Iodothyronine Deiodinases in Different Areas." *Journal of Clinical Endocrinology & Metabolism*; 89 (2004): 3117–28.

Epilogue: A Sting in the Tail

1. Locke, Michael. "Professor Sir Vincent B. Wigglesworth: Biography and Contributions to Insect Morphology and Embryology for the 1992 Distinguished International Award." *International Journal of Insect Morphology & Embryology*; 21 (1992): 101–15.

2. Locke was replying to my question in a personal communication.

3. Edwards, John S. "Sir Vincent Wigglesworth and the Coming of Age of Insect Development." *International Journal of Developmental Biology* 42 (1998): 471–3. See also, Edwards, John S. "In Memoriam: Sir Vincent Brian Wigglesworth (1899–1994)." *Developmental Biology* 166, 2 (1994): vi-362.

4. Lawrence, Peter A., and Michael Locke. "A Man for Our Season." *Nature* 386 (1997): 757–8.

5. Akman, Leyla, Atsushi Yamashita, et al.. "Genome Sequence of the Endocellular Obligate Symbiont of Tsetse Flies, *Wigglesworthia glossinidia*." *Nature Genetics* 32 (2002): 402–7.

6. Kuhn, Thomas S. *The Structure of Scientific Revolutions*, 3rd ed. (Chicago: University of Chicago Press, 1996), p. 1.

7. Gould, Stephen Jay. *Eight Little Piggies* (London: Jonathan Cape, 1994), p. 146.

8. See Williamson, Donald I. *The Origins of Larvae* (Dordrecht and Boston: Kluwer Academic Publishers, 2003), ch. 14 and fig 14.2d.

9. Williamson, Donald I., and A. L. Rice. "Larval Evolution in the Crustacea." *Crustaceana* 69, 3 (1996): 267–87.

10. Williamson, Donald I. "Caterpillars Evolved from Onychophorans by Hybridogenesis." *PNAS* (*Proceedings of the National Academy of Sciences*) 106 (2009): 19901–5.

11. Syvanen, Michael, and Clarence I. Kado, eds. *Horizontal Gene Transfer* (London: Chapman & Hall, 1998).

12. Gould, Stephen Jay. *Eight Little Piggies*, p. 73.

13. Gould, Stephen Jay. *Wonderful Life: The Burgess Shale and the Nature of History* (London: Penguin, 1991), pp. 23–4.

14. The actual dating of time periods around the junction of the Proterozoic and Cambrian is variable from one authority to another. Some, for example, regard the Cambrian as dating from 570 million years ago, so the Burgess Shale is Middle Cambrian, while others regard it as 545 or 540 million years ago. The only safe measure is to qualify such designations with the actual years.

15. See illustrated Vendian animals at http://www.ucmp.berkeley.edu/vendian/critters.html. There is also a very interesting and helpful site on the Ediacaran Assemblage at www.peripatus.gen.nz/paleontology/Ediacara.hml.

16. Gehling, J.G. "Earliest Known Echinoderm: A New Ediacaran Fossil from the Pound Subgroup of South Australia." *Alcheringa* 11 (1987): 337–45.

17. Morris, Simon Conway. "The Cambrian 'Explosion': Slow-fuse or Mega-tonnage?" *Proceedings of the National Academy of Sciences* 97 (2000): 4426–9.

18. Babcock, Loren E., Wentang Zhang, and Stephen A. Leslie. "The Chengjian Biota: Record of the Early Cambrian Diversification of Life and Clues to Exceptional Preservation of Fossils. *GSA Today* (February 2001), http://www.geosociety.org/gsatoday/archive/11/2/pdf/i1052-5173-11-2-4.pdf.

19. Darwin, Charles. *On the Origin of Species by Means of Natural Selection* (London: John Murray, 1859), ch. 9.

20. Gould, Stephen Jay. *Wonderful Life*, p. 14.

21. Morris, Simon Conway. *The Crucible of Creation: The Burgess Shale and the Rise of Animals* (Oxford and New York: Oxford University Press, 1998), pp. 104–6; 131. Shu, D-G., H-L. Luo, et al. "Lower Cambrian Vertebrates from South China." *Nature* 402 (1999): 42–6. See also Valentine, James W. *On the Origin of Phyla* (Chicago: University of Chicago Press, 2004), p. 182 and fig. 5.18.

22. Whittington, Harry B. "Redescription of *Marrella splendens* (Trilobitoidea) from the Burgess Shale, Middle Cambrian, British Columbia." *Geological Survey of Canada Bulletin* 209 (1971): 1–24.

23. Hughes, C.P. "Redescription of *Burgessia bella* from the Middle Cambrian Burgess Shale, British Columbia." *Fossils and Strata* (Oslo) 4 (1975): 415–35. There are some beautiful pictures of this creature and many others at Stefan Bengston's site: http://luna.geol.niu.edu/2000_1/fossils/text.pdf.

24. Gould, Stephen Jay. *Wonderful Life*, p. 138.

25. Morris, Simon Conway. "*Nectocaris pteryx*, a New Organism from the Middle Cambrian, Burgess Shale of British Columbia." *Neues Jahrbuch für Geologie und Paläontologie* 12 (1976): 705–13.

26. Morris, Simon Conway. *Crucible of Creation*, pp. 109–10.

27. Gould, Stephen Jay. *Wonderful Life*, p. 146.

28. Ibid., p. 160.

29. Valentine, James W. *On the Origin of Phyla*, p. 189, *et seq.*

30. Willmer, Pat. *Invertebrate Relationships: Patterns in Animal Evolution* (New York and Cambridge: Cambridge University Press, 1990), p. 332.

31. The ostensible theme of the conference was "The Human Brain in the Context of Natural History: 3000 Million Years of Evolution of Sensory Systems." It was held at the Rockefeller Foundation's Study and Conference Center. This was Williamson's first public address in person since his stroke. He was assisted in getting there by Robert Sternberg and he worried about his ability to present himself. In the event, he received an enthusiastic reception. The printed report of the conference is currently in press. See, Vickers, Sonya E., and Donald I. Williamson. "Larval Transfer: Evolution by

Hybridization." In Margulis, Lynn, C. Asikainen, and W. Krumbein, eds. *Chimera and Consciousness: Evolution of the Sensory Self* (White River Junction, Vt.: Chelsea Green, in press).

32.  Williamson's essay, along with others, was subsequently published in Brockman, John, ed. *What We Believe But Cannot Prove: Today's Leading Thinkers on Science in the Age of Certainty* (London: Pocket Books, 2006).

33.  Valentine. James W. *On the Origin of Phyla*, p. 179, *et seq.*

34.  Williamson, Donald I. "Hybridization in the Evolution of Animal Form and Life-cycle." *Zoological Journal of the Linnean Society*. 148 (2006): 585–602.

35.  Morris, Simon Conway. *Crucible of Creation*, p. 159, *et seq.* Valentine, James W. "Adaptive Strategies and the Origins of Grades and Ground Plans." *American Zoologist* 15 (1975): 391–404.

36.  Williamson, Donald I. "Larval Genome Transfer: Hybridogenesis in Animal Phylogeny." *Symbiosis* (in press, 2011).

37.  Syvanen, Michael, and Jonathan Ducore. "Whole Genome Comparisons Reveals a Possible Chimeric Origin for a Major Metazoan Assemblage." *Journal of Biological Systems* 18 (2010): 261–75.

# Acknowledgements

To a scientist dedicated to the search for objective truth, ideas are precious things, never more so than in the intellectual contribution any one individual has made to the ever-metamorphosing wealth of knowledge that is the cherished common legacy of a discipline. To research a book such as this, one inevitably invades these domains, sometimes sharing, at other times questioning, the inspirations behind a lifetime of effort. My debts are many, for this book is an essential part of a sequence of other books, with all that I owe to those who helped me. Thus it is with humility and the utmost gratitude that I express my thanks for the generosity of time and spirit with which so many busy colleagues have welcomed my intrusion into their lives and work.

My special thanks, inevitably, go to Donald Williamson, of the Marine Laboratory, Isle of Man, and to his wife and family for their hospitality; to Professor Jonathan Wigglesworth, for the interview and documents concerning his father, Vincent, and for the hospitality shown to me by him and his wife; to Professor Lynn Margulis, University of Massachusetts, Amherst, to whom I owe so much already, and whose original suggestion led to my discovering Don Williamson's extraordinary line of experiments; to Professor Chris Curtis, London School of Hygiene and Tropical Medicine; to Professor Simon Maddrell, Department of Zoology, University of Cambridge, to Professors

James W. Truman and Lynn M. Riddiford, both of the Department of Biology, University of Washington; to Professor Luis Villarreal, Department of Molecular Biology and Biochemistry and Director of the Center for Virus Research, University of California, Irvine; and to Professor R. Thomas Zoeller, Biology Department, Morrill Science Center, University of Massachusetts, Amherst.

I am also deeply indebted to the following, who were generous with their time and assistance: Michael L. Arnold, Associate Professor, Department of Genetics, University of Georgia; Henry Bennett-Clark, Zoology Department, University of Oxford; Professor David Bradley, Department of Infectious and Tropical Diseases, London School of Hygiene and Tropical Medicine; Professor Donald D. Brown, Department of Embryology, Carnegie Institution, Washington, DC; Dr Alan J. Chalk, University of Delaware, for the interesting exchange of ideas; David Champlain, Assistant Professor, Department of Biological Sciences, University of Maine; Jonathan D. Cowart, Department of Biology and Marine Biology, University of North Carolina at Wilmington; R. Henry Disney, Department of Zoology, University of Cambridge; Brian Furner, Librarian and Director of Information Service, London School of Hygiene and Tropical Medicine; Professor Michael G. Hadfield, Kewalo Marine Laboratory, University of Hawaii at Manoa; Michael Hart, Associate Professor, Department of Biological Sciences, Simon Fraser University, British Columbia; Sebastian P. Holmes, University of Liverpool; Michael Locke, Professor Emeritus, Department of Biology, University of Western Ontario; John M. McCoy, Vice President, Discovery Biology, Biogen Idec Inc; François Mallet, Ecole Normale Superieure de Lyon; Dr Claus Nielsen, Associate Professor and Curator, Invertebrate Department, Zoological Museum, Copenhagen; Mark G. Nielsen, Assistant Professor of Biology, University of Dayton, Ohio; Professor Trevor Norton, Department of Marine

Biology, University of Liverpool; Rachel O'Neill, Assistant Professor, University of Connecticut; Professor Sidney K. Pierce, Professor and Chair, Biology Department, University of South Florida; Loren Rieseberg, Professor, Biology Department, Indiana University; the late Professor Sir Richard Southwood, Zoology Department, University of Oxford; Linda Van Speybroeck, FWO post-doctoral researcher, Ghent University; Professor Richard Strathmann, Friday Harbor Laboratories, San Juan Island, Washington; Billie J. Swalla, Associate Professor, Biology Department, University of Washington; R.B. Toms, Department of General Entomology, Transvaal Museum, Pretoria, South Africa; J. Carel von Vaupel Klein, Leiden University; Kaori Wakabayashi, Department of Biology, Toyama University; Graham B. White, University of Florida; Cheryl Whitehorn, Medical Entomology Technician, London School of Hygiene and Tropical Medicine; James B. Whitfield, Department of Entomology, University of Illinois. I would also like to thank the archives department at Caius College, University of Cambridge, and the Harvard University Archives. I am, moreover, indebted, if not in person but in inspiration to the work and writings of Sir Vincent B. Wigglesworth, Charles Darwin, and Jean-Henri Fabre.

It gives me the greatest pleasure to acknowledge the contribution of Joni Praded at Chelsea Green and Robin Dennis at Oneworld, whose belief in and dedication to this book was unshakeable. And what a wonderful opportunity it was to work with Jonathan Cobb, whose insightful editorial comments and suggestions made a very real contribution to the final text. My agent, Jonathan Pegg, was a pillar of support, as were the project manager, Pati Stone, and the copyeditor, Cannon Labrie, and all others at Chelsea Green. It was a real pleasure to work with such an efficient group.

As always I acknowledge the love and support of my wife, Barbara, invariably my toughest critic and true believer.

# Index